Textbooks in Electrical and Electronic Engineering

Series Editors

G. Lancaster   E. W. Williams

1. Introduction to fields and circuits
   GORDON LANCASTER

2. The CD-ROM and optical recording systems
   E. W. WILLIAMS

3. Engineering electromagnetism: physical processes and computation
   P. HAMMOND and J. K. SYKULSKI

Engineering Electromagnetism

Physical Processes and

Computation

■

P. Hammond
and
J. K. Sykulski

Department of Electrical Engineering
University of Southampton

OXFORD   NEW YORK   TORONTO
OXFORD UNIVERSITY PRESS

Oxford University Press, Walton Street, Oxford OX2 6DP

Oxford New York
Athens Auckland Bangkok Bombay
Calcutta Cape Town Dar es Salaam Delhi
Florence Hong Kong Istanbul Karachi
Kuala Lumpur Madras Madrid Melbourne
Mexico City Nairobi Paris Singapore
Taipei Tokyo Toronto
and associated companies in
Berlin Ibadan

Oxford is a trade mark of Oxford University Press

Published in the United States
by Oxford University Press Inc., New York

A catalogue record for this book is available from the British Library

Library of Congress Cataloging in Publication Data
P. Hammond and J. K. Sykulski.
Engineering electromagnetism: physical processes and computation
(Textbooks in electrical and electronic engineering; 3)
1. Electromagnetic theory. 2. Electronics. 3. Electric
engineering. I. Sykulski, J. K. II. Title. III. Series.
QC760.H324   1994   621.3—dc20   92–50060
ISBN 0 19 856289 6 (Hbk)
ISBN 0 19 856288 8 (Pbk)

Printed and bound in Great Britain by
Biddles Ltd, Guildford and King's Lynn

Preface

A working knowledge of electromagnetism is an essential part of the intellectual equipment of every electrical or electronic engineer. However, such knowledge is not easily acquired because of the inherent difficulties of the subject. Engineering is generally concerned with material objects and processes, but electromagnetic effects exist in empty space as well as in matter. Considerable imaginative powers are needed to relate such effects to more common-sense phenomena.

The object of this book is first to stimulate the reader's imagination by helping him to acquire a vocabulary of electromagnetic terms which is related by analogy to more familiar processes. Much of the content of the first few chapters is based on the idea of flux as an analogy of fluid flow and of potential difference as an analogy of force and pressure. The circuit parameters of resistance, capacitance, and inductance are presented in terms of energy dissipation, potential energy, and kinetic energy. Time-varying effects are related to energy interchange between potential and kinetic energy. Local field effects are approached by dividing the flux into small tubes, and the potential difference into small slices of space.

Once the vocabulary has been acquired the language of electromagnetism can be learned by its use in engineering applications. The second aim of the work is to guide the reader in the use of electromagnetic terms and relationships. We use the notion of tubes and slices not only as a conceptual tool, but also as the basis of numerical computation. Many books are based chiefly on analytical formulae, which are, of course, extremely powerful. But such formulae are simple only when they describe extremely simple geometrical shapes such as flat plates of infinite extent or infinitely long cylinders of constant cross-section. The numerical method of tubes and slices is not restricted in this way. For its implementation it needs a computer program, and this program is provided in the disk accompanying the book.

In engineering practice, electromagnetic problems are invariably solved by numerical computation. It is, therefore, desirable that the student should become familiar with such methods as early as possible. If he acquires skill by using a very simple, but accurate, method such as that of tubes and slices, he will be able to use more complicated methods later without being overawed by their complexity or accepting their results uncritically. A comprehensive introduction to a number of analytical, numerical, and analogue methods of field computation is provided in the last section of this book. As most of the commercially available electromagnetic software is based on the finite element method, this technique is discussed in more detail. The emphasis is on understanding and practical applications with the mathematical formulations presented in the simplest possible way.

The program supplied with this book is suitable for interactive use on personal computers. The time of computation is so short that the numerical results appear almost instantaneously. The user can easily make alterations to the geometry of the system which he is studying and can therefore carry out simple design and optimization procedures.

We have used the method of this book, and the accompanying disk, for several years and have found that undergraduates take to it like ducks to water. This applies even to students who regard electromagnetism as a secondary interest. We have also used the method successfully on courses for mature students who had found a need for acquiring a knowledge of electromagnetism.

We hope the reader will enjoy the book and let us know of any suggested changes which would improve it.

Southampton   P.H.
August 1993   J.K.S.

Contents

1 Flow of steady current, resistors

1.1	Flow of steady current	1
1.2	Resistance	2
1.3	Tubes and slices	3
1.4	Computation of steady-current flow	7
1.5	More calculations using tubes and slices	11
1.6	The geometrical structure of fields	13
	Exercises	13

2 Electrostatic fields, capacitors

2.1	Energy storage	17
2.2	Electric flux	19
2.3	Tubes and slices in the electrostatic field	22
2.4	Calculation of capacitance using tubes and slices	23
2.5	Parallel-plate capacitors with fringing	26
2.6	System energy	26
2.7	The charge distribution in conductors	29
2.8	Insulators	30
2.9	The effects of polarization	31
2.10	Partial and total fields	32
2.11	Boundary conditions	33
	Exercises	34

3 Steady currents and magnetostatics

3.1	Electrokinetic energy	40
3.2	Electricity and magnetism	43
3.3	Tubes and slices, inductance, and boundary conditions	45
3.4	Magnetic shells	48
3.5	Magnetomotive force	49
3.6	The magnetic circuit and permeance	51

3.7 Calculation of permenance using tubes and slices   54
3.8 The magnetic field of a long cylindrical conductor   56
3.9 The magnetic field of a current element   58
3.10 The force on a current in a magnetic field   61
3.11 Stress in a magnetic field   62
Exercises   65

4 Electric and magnetic fields as vectors

4.1 Introduction to vectors   70
4.2 The gradient vector   71
4.3 The flux vector   73
4.4 Laplace's and Poisson's equations   78
4.5 Polar and axial vectors   81
4.6 Vortex fields   82
4.7 The independence of conservative and vortex fields   84
4.8 The magnetic vector potential   86
4.9 The uniqueness of vector fields   87
Exercises   89

5 Electromagnetic induction

5.1 The importance of time   92
5.2 Motional electromotive force and electromechanical
energy conversion   94
5.3 The motion of charge in free space   100
5.4 Faraday's law   101
5.5 The transformer   105
5.6 Self-inductance and mutual inductance   107
5.7 The skin effect and eddy currents   109
5.8 Electrokinetic momentum and the vector potential   117
Exercises   119

6 Electromagnetic radiation

6.1 Displacement current and Maxwell's equations   123
6.2 Electromagnetic waves   125
6.3 Retarded potentials   131
6.4 The field of an oscillating doublet or electric dipole   134
6.5 Radiation from a linear antenna   139
6.6 Directivity of antenna arrays   141

6.7 Waves guided by conductors   143
6.8 Waveguide modes   146
Exercises   149

7 Computation of fields

7.1 Introduction   151
7.2 Separation of variables   152
7.3 Separation of variables in electrostatics   155
7.4 Magnetostatic screening   158
7.5 The method of images   161
7.6 Analogue methods   164
7.7 Tubes and slices   167
7.8 The finite-difference method   169
7.9 The finite-element method   174
7.10 Discretization and matrix assembly   179
7.11 Solving the system equations   183
7.12 Boundary elements   186
7.13 Tubes and slices and finite elements   187
7.14 The CAD environment   188
Exercises   192

8 Engineering applications

8.1 Introduction   199
8.2 An inductive sensor: a case study   200
8.3 The rotating field   210
8.4 Torque in rotating machines   215
8.5 Energy storage and forces in magnetic-field systems 218
8.6 Magnetic materials   221

Appendices

A1 TAS tutorial   227
A2 Useful mathematical formulae   243
A3 Physical and materials constants   247
A4 The SI system   248
A5 Bibliography   249
A6 A summary of electromagnetic relationships   251

Index   253

For

Hanna and Adam

Flow of steady current, resistors 1

1.1 Flow of steady current

The heating elements of electric space or water heaters are familiar objects. Normally such devices are supplied with electricity of a power frequency of 50 or 60 Hz, but in our initial discussions we want to avoid any complications associated with alternating currents, so we shall assume steady or direct current. Later we shall find that in practical heaters the time variation is not important.

How shall we describe the electrical behaviour of the heating element? Its use is to convert electrical energy into heat energy, so we shall need to consider the electrical energy. This energy has two aspects. First, there is the flow of electric current, which is measured in amperes, and secondly there is the potential difference between the ends of the element, which is measured in volts. The product of volts and amperes is the rate at which energy is converted. For steady current this rate is constant, so that the energy can be obtained by multiplying the rate by the time. The energy depends both on the electric supply and on the construction of the heating element. It is helpful to separate these two factors and to isolate the properties of the heater, which can of course be operated from different supplies. It is found experimentally that the current and voltage are generally proportional to each other, so that their ratio is independent of the supply and depends only on the heater. We can write

(1.1)
$$\frac{V}{I} = R,$$

where R is the resistance of the element in ohms. For metallic conductors, R is independent of V and I, but it does depend on the temperature. Such conductors are said to obey Ohm's law. We represent the heating element by the circuit diagram of Fig. 1.1.

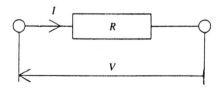

Fig. 1.1 A resistor.

No doubt all this is very familiar to the reader, who may by now be wondering if this work is written for beginners rather than for mature students. If the discussion was left at this stage, it would indeed be elementary. But now we want to ask some awkward questions to show the limitation of this circuit representation.

1.2 Resistance

What does the property R tell us? It is the ratio of two external measurements, which treat the heater as a single entity. It does not tell us anything about the inside of the element. Given a certain heater we can find its resistance, but we cannot design a heater. Before that can be done we need to know what is happening inside the element, and such knowledge will be achieved only if the principles of the flow of electricity are understood, and if these principles can be organized into a numerical form suitable for computation. That is the type of problem which is addressed by this book.

Consider first the physical mechanism of current flow. It is known that conduction in metals is due to the motion of electrons which have sufficient energy to move freely through the material. Inside the metal there is no net electric charge, so that the current consists of a relative displacement of the moving negative electrons through the positively charged lattice. The resistance is due to collisions between the electrons and the lattice. These cause lattice vibrations and therefore generate heat. A force is required to keep the current flowing. Ultimately this force is due to the potential difference between the ends of the element. But this does not explain how, or by what means, this force is distributed. Experiments with parallel-sided conductors of different cross-sectional areas and lengths, but of the same material, show that

(1.2)
$$R = \frac{l}{\sigma S},$$

where l is the length, S is the cross-sectional area and σ is the conductivity. This shows that, as expected, the force is distributed uniformly along the length and that the current is uniform across the section of such conductors. The local form of Ohm's law can therefore be written as

(1.3)
$$\sigma E = J,$$

where E is the potential gradient, V/l, and J is the current density, I/S. The current density can be written as

(1.4)
$$J = -nev,$$

where n is the number of free electrons per unit volume, e is the electronic charge, and v is the average drift velocity. Equation (1.3) can now be interpreted as the statement that the local force is proportional to the local drift velocity, and this explains the mechanism of eqn (1.1). We are now in a position to design conductors with a particular resistance using eqn (1.2). However, this equation holds only if the conductor has a constant cross-sectional area. We cannot yet design conductors such as those shown in Fig. 1.2. Of course we can measure their resistance and modify the shape by trial and error. But that would be expensive and unsatisfactory.

Fig. 1.2 Examples of conductors.

1.3 Tubes and slices

To examine the problem of the distribution of the current and the voltage, consider conductors in the form of a block of constant thickness. First consider the block shown in Fig. 1.3.

Let us assume that the current enters the block perpendicularly at the shaded end and leaves in a similar manner on the other side. Since Ohm's law (eqn 1.3) requires that there has to be a force driving the current, it can be deduced that there is no force along the shaded side and therefore that this side must be an equipotential surface. Also, since no current enters or leaves except at the ends, all the other surfaces lie along the direction of flow. The boundary conditions are therefore that the block has two equipotential surfaces and four impermeable surfaces. Notice that equipotential surfaces are always perpendicular to the current flow. The external connections could be made by two highly conducting sheets for the ends and by an insulating sheet wrapped round the other four surfaces.

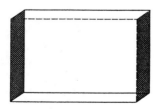

Fig. 1.3 A conducting block.

In conductors of arbitrary shape, we shall find it useful to introduce a subdivision of the conductor. Our discussion so far suggests two types of subdivision. We could regard a single conductor as a set of conductors in parallel. We shall call this a subdivision into tubes of current. Or we could regard the single block as a set of conductors in series. We shall call this a subdivision into slices of potential difference. The internal boundaries for the tube connection of Fig. 1.4 would be very thin insulating sleeves and the internal boundaries of the slice connection of Fig. 1.5 would be very thin sheets of infite conductivity. The total resistance, R, of the block in terms of the resistances of the individual resistors, r_t, would be given by

(1.5)
$$\frac{1}{R} = \sum \frac{1}{r_t}$$

for the tube connection and by

(1.6)
$$R = \sum r_s$$

for the slice connection. All the parallel conductors have the same potential difference between their ends, and all the series conductors carry the same current.

Since the direction of the current density is known everywhere, we know how to insert the insulating sleeves in Fig. 1.4 and the conducting sheets in Fig. 1.5 so as to leave the flow undisturbed. Thus, if there are m tubes of equal cross-sectional area eqn (1.2) can be used to write

(1.7)
$$r_t = \frac{l}{\sigma S/m} = \frac{ml}{\sigma S}.$$

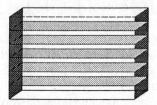

Fig. 1.4 Tubes of current.

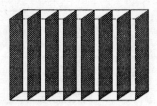

Fig. 1.5 Equipotential slices.

This satisfies eqn (1.5), since

(1.8)
$$\frac{1}{R} = \frac{\sigma S}{l} = \frac{m}{r_t}.$$

Similarly, if there are n slices,

(1.9)
$$r_s = \frac{l/n}{\sigma S}$$

and

(1.10)
$$R = nr_s.$$

In this example nothing has been gained by the subdivision of the conductor into tubes and slices, because the shape of the tubes and slices is exactly the same as that of the original block. Geometrically we can use the same coordinates for the local directions of the current and the voltage as for the global directions on the surfaces of the block. The power of our method of tubes and slices becomes apparent only if the shape of the conductor is more complicated. Consider the resistance of the block in Fig. 1.6. Here the local direction of the current is known only on the surfaces. Similarly the direction of the equipotentials is known only on the surfaces. Equation (1.3) is still correct, but it cannot be used to calculate the resistance directly. Equation (1.2) cannot be used, because both l and S vary. We do, however, know that everywhere the tubes and the slices must be at right angles, because there is no flow along an equipotential. This knowledge can be used to sketch the flow, as in Fig. 1.7. Each part of the

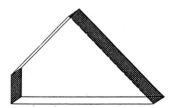

Fig. 1.6 A quadrilateral block.

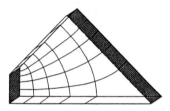

Fig. 1.7 Tubes and slices.

subdivided system has curved surfaces; but as the number of subdivisions into tubes and slices is increased the curvature becomes less, and in the limit the surfaces are flat. With extensive subdivision we can therefore arrive at a system of rectangular blocks for which eqn (1.2) holds.

Suppose there are m tubes and n slices. Let the length of a typical small block be l_j and let the cross-sectional area be S_i. Clearly these variables will vary from piece to piece. For a single tube the resistance will be

(1.11)
$$r_t = \sum_{j=1}^{n} \frac{l_j}{\sigma S_i}.$$

For the whole block, all the tubes must be connected in parallel, so that

(1.12)
$$\frac{1}{R} = \sum_{i=1}^{m} \left(\sum_{j=1}^{n} \frac{l_j}{\sigma S_i} \right)^{-1},$$

and finally

(1.13)
$$R = \left[\sum_{i=1}^{m} \left(\sum_{j=1}^{n} \frac{l_j}{\sigma S_i} \right)^{-1} \right]^{-1}.$$

Alternatively,

(1.14)
$$\frac{1}{r_s} = \sum_{i=1}^{m} \frac{\sigma S_i}{l_j}$$

and

(1.15)
$$R = \sum_{j=1}^{n} \left(\sum_{i=1}^{m} \frac{\sigma S_i}{l_j} \right)^{-1}.$$

Equations (1.13) and (1.15) give the same result, if the subdivision is sufficiently small. Further, if the division is such that for every piece $S_i = K l_i$, then

(1.16)
$$R = \frac{n}{m} \frac{1}{\sigma K}.$$

For two-dimensional flow, the area S becomes a line. It is then possible to put $K = 1$ by making 'curvilinear squares'.

We shall investigate the numerical procedures involved in such calculations. But before we do so another interesting question presents itself. The subdivision into rectangular blocks can never be totally correct, because it would require an infinite number of elements. A finite number of tubes and slices will always disturb the flow. Will this process increase the resistance or decrease it?

This question was answered by James Clerk Maxwell in his famous book *Electricity and Magnetism*. His argument was based on the physical consideration of current flow and it is beautifully simple. Suppose we subdivide into tubes only. This can be done by inserting thin insulating sleeves between the tubes. Maxwell argued that the insertion of insulating material must always increase the resistance unless the flow is undisturbed. Very thin sleeves which are in the correct direction everywhere will have a negligible effect, but if such sleeves are not exactly in the direction of the current flow they will increase the resistance. The undisturbed resistance is therefore a minimum.

Equally, the insertion of infinitely conducting sheets for slices will reduce the resistance if they disturb the flow. The undisturbed resistance is therefore a maximum.

Thus a division into tubes and slices enables upper and lower bounds to be calculated for the unknown resistance. The average of these two bounds often provides a highly accurate estimate, and this method avoids the need for very small subdivisions.

1.4 Computation of steady-current flow

We will now proceed to look at some numerical examples. As a matter of convenience, and without sacrificing any generality of formulation, let us concentrate on current flow in two dimensions, which is easier to visualize. This assumption simply means that our results will be calculated per unit depth, giving the resistance in units of $\Omega\,\mathrm{m}^{-1}$.

Let us investigate ways of finding the approximate resistance of the piece of material shown in Fig. 1.8, having a conductivity σ (or a resistivity $\rho = 1/\sigma$), which is between a pair of vertical electrodes.

First of all, it is rather surprising and interesting that this simple shape poses quite a challenging computational problem, usually calling for a complicated field solution, and yet there is only a slight departure from a simple rectangular shape for which an answer could be given instantly.

Let us make a guess (guided by the solution for a rectangle) about the shape of the equipotential lines (slices) inside the region and assume them to

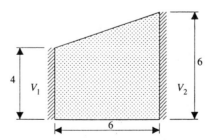

Fig. 1.8 Example.

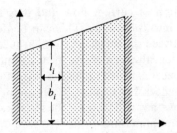

Fig. 1.9 Slices.

be straight vertical lines as shown in Fig. 1.9. We know in advance, from the previous discussion, that the consequence of such a step will be to give a lower bound for the resistance (insertion of highly conductive thin sheets). We have used this step because it produces a system consisting of simple, nearly rectangular blocks which are connected in series. It is true that these blocks are not perfect rectangles, but with $b_i \gg l_i$ the approximation of using a mean width, b_i, can be justified. The resistance of a segment i will be

$$R_i = \frac{l_i}{b_i}\,\rho,$$

and the total resistance will be

$$R = \sum_{i=1}^{n} R_i.$$

For two slices $(n=2)$

$$R_- = \frac{3}{4.5}\,\rho + \frac{3}{5.5}\,\rho = 1.212\rho,$$

where the subscript minus sign indicates the lower bound. For ten slices $(n=10)$ we would probably use a programmable calculator or a small computer, to find that

$$R_- = 1.216\rho,$$

which is marginally better than the previous result (it is indeed better because it is a higher value for the lower bound).

We now turn our attention to 'tubes' of current. We intend to specify a distribution of tubes in which the calculations will be easy. We know that by doing so we can expect an upper bound for the resistance. Let us first make a rather crude approximation by assuming that all the current leaving the left electrode will flow horizontally to the right electrode, effectively ignoring the presence of conductive material at the top (the shaded area in Fig. 1.10). We may be accused of being rather extreme in our approach, but

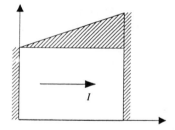

Fig. 1.10 A simple tube.

all we have actually done is to make a somewhat violent repositioning of a tube boundary, consequently increasing the value of an upper bound. The resistance of the perturbed system may be found easily as

$$R_+ = \tfrac{6}{4}\rho = 1.5\rho,$$

where the subscript plus sign indicates an upper bound. The average of the two bounds equals

$$R = \tfrac{1}{2}(R_+ + R_-) = \tfrac{1}{2}(1.5\rho + 1.216\rho) = 1.358\rho \pm 10.5\%.$$

Note that both bounds, and hence the ± 10.5 per cent accuracy, are guaranteed. Let us try to improve the accuracy of our result.

The upper-bound calculation, based on the current flow assumed in Fig. 1.10, certainly leaves scope for improvement. A possible improved scheme is illustrated in Fig. 1.11. First, a diagonal line is constructed (the broken line). Then a number of tube lines are drawn parallel to one side of the quadrilateral up to the diagonal and parallel to the opposite side beyond the diagonal.

A system of tubes is obtained, each containing two pieces connected in series. The resistance of one of the tubes will be given by

$$R_i = \frac{l_{1i}}{b_{1i}}\rho + \frac{l_{2i}}{b_{2i}}\rho.$$

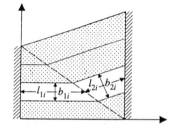

Fig. 1.11 Alternative tubes.

This should be a good approximation since, normally, $l_i \gg b_i$. Finally, all tubes are connected in parallel, so that

$$\frac{1}{R_+} = \sum \frac{1}{R_i}.$$

The calculations involved are slightly more complicated, as the mean length and width of every piece has to be found. Hand calculations may be tedious, but even a small personal computer will perform them in a fraction of a second. As an example, a subdivision into eight tubes gives an upper bound for the resistance of $R_+ = 1.296\rho$. When combined with the previously calculated lower bound this gives

$$R = \frac{R_+ + R_-}{2} = 1.256\rho \pm 3.2\%.$$

If the two assumed distributions of tubes and slices are combined (Fig. 1.12) the reasons why the two bounds are not equal in our calculations can be seen. In an accurate solution, the tube and slice lines would be orthogonal everywhere. This requirement has been relaxed and thus a system is obtained which is only approximately orthogonal. Hence our value of the resistance is estimated within a certain error band, given by the guaranteed upper and lower bounds. The actual error is more difficult to assess. In order to get some idea of its value compare our result with an independent solution obtained using a *finite-element* method (for a discussion of this and other methods of field solutions see Chapter 7). Using a very fine mesh and a 'dual formulation' (to reduce errors) the value $R = 1.243\rho$ is obtained. Comparing the two results it can be seen that, although the value of 1.256ρ is guaranteed to be accurate within ± 3.2 per cent, the actual accuracy is even better (around 1 per cent). The reason for this remarkable improvement in accuracy is that the errors involved in calculating the dual bounds largely cancel out.

The closeness of the two bounds, together with the orthogonality (or lack of it) of the tube and slice lines, form a criterion for possible further improvements in the assumed distributions.

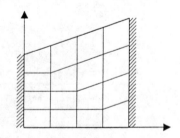

Fig. 1.12 Tubes and slices.

1.5 More calculations using tubes and slices

We can now embark on a slightly more ambitious project. Let us calculate
the resistance for the system shown in Fig. 1.13(a). We shall employ the
technique used in the previous example (and illustrated in Fig. 1.11), for
both the system of tubes and of slices. We start by drawing a few
construction lines and thus defining a system of quadrilaterals with
diagonals, as shown in Fig. 1.13(b). There is no need to worry too much
about the precise position of the construction lines. If this step is performed
using a computer with the aid of interactive graphics, there will be an
opportunity to make small corrections later on. At the moment we satisfy
ourselves with the fact that we have some basic structure which enables us
to draw tubes and slices in a familiar fashion. That is, lines are drawn
parallel to one or other of the sides of each quadrilateral taking account of
the position of the diagonals. We shall arrive at distributions similar to
those illustrated in Fig. 1.13(c) and (d) depending on the number of
subdivisions used. The final set of tubes/slices will look like that shown in
Fig. 1.13(e).

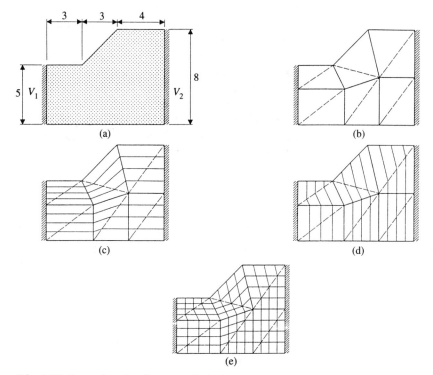

Fig. 1.13 Examples of resistance calculation.

The calculation procedure is very simple in principle. We have a number of tubes (eight in Fig. 1.13(c)), each formed by six simply shaped pieces connected in series. The eight tubes are then connected in parallel. Similarly, each of the twelve slices in Fig. 1.13(d) is made up from four small elements connected in parallel. All the slices are then connected in series. Thus all we need in both the upper- and lower-bound calculations is to add component resistances as a series/parallel or parallel/series connection (see eqns 1.13 and 1.15). Calculations like these can be made in a fraction of a second on a personal computer. A system like that shown in Fig. 1.13(e) will give the following dual bounds

$$R_+ = 1.706\rho$$
$$R_- = 1.595\rho,$$

hence

$$R = \tfrac{1}{2}(R_+ + R_-) = 1.65\rho \pm 3.4\%.$$

Calculations similar to those described above are frequently performed as part of a design process. The designer may find the guaranteed accuracy unsatisfactory, in which case ways of improving the results by making the two systems of tubes and slices more orthogonal would be sought. Some changes – such as repositioning the construction lines, choosing an alternative diagonal, or adding new lines – can be easily accommodated in a computer program using interactive graphics.

Finally, for purposes of comparison, let us solve the example shown in Fig. 1.13(a) using a different method, such as the finite-element method. A vaue of $R = 1.652\rho$ is obtained, showing that our tubes-and-slices estimate is only 0.12 per cent different from this value. The field plot obtained from the finite-element solution is shown in Fig. 1.14.

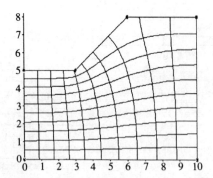

Fig. 1.14 A field plot of the finite-element solution.

1.6 The geometrical structure of fields

Some readers will have realized by now the potential of the approximate solutions offered by geometrical methods such as the tubes-and-slices approach. The benefits include speed and economy of computation, simplicity of formulation, guaranteed bounds of solution, ease of data preparation (preprocessing) and results retrieval (postprocessing) and strict control over the whole process. It will also have been noticed that, even without a special computer program, good use can be made of dual-bound approximations by using the cruder simplifications appropriate to hand calculations, which can still give very good answers. We shall frequently refer to this powerful technique throughout the text of the book.

The good results obtained using the tubes-and-slices approach are, of course, not accidental but are due to the physical and mathematical foundation of the method. A full treatment is beyond the scope of this book, and the interested reader is referred to various publications in professional journals (see the Bibliography in Appendix 5).

Tubes of flow, and slices of equipotential surfaces, are inherent properties of vector fields and are very helpful in visualizing field distributions, even if a different method of calculation is used from that used here. The result of all kinds of calculations are generally displayed in the form of field plots. It is not surprising, therefore, that these tubes and slices form a sound basis for actual solutions.

Exercises

1.1 Give a general account of Ohm's law, and state what is meant by resistance.

1.2 What is the resistance of a parallel-sided block of conducting material of length, l, cross-sectional area, S, and conductivity, σ?

1.3 Explain what is meant by a tube of current. What material could be used physically to model the tube boundary? How will this material affect the flow of current; and, thus, what will happen to the value of the resistance?

1.4 Explain what is meant by a slice of equipotential. Suggest an appropriate material physically to simulate a slice. How is the flow of current affected, and what happens to the resistance, when the slice disturbs the current flow?

1.5 Discuss the use of *tubes and slices* in calculations of dual bounds of a direct current (d.c.) resistance. Demonstrate how the value of the resistance could be estimated as a series/parallel connection of simply shaped component resistors. Discuss the computational advantages of such a scheme.

1.6 Discuss sources of errors in calculations based on the *tubes-and-slices* approach and suggest ways of improving the solution. What criteria could be used when modifying the distributions to achieve an improved accuracy?

1.7 Using the idea of *tubes and slices*, estimate the upper and lower bounds of a d.c. resistance (per unit length) between the pair of vertical electrodes shown in Fig. 1.15 and compare your result with the value obtained from a finite-element program, $R = 7.6\rho \ \Omega \ \mathrm{m}^{-1}$.

Answer

$$R_+ = \frac{1+5+4}{1}\rho = 10\rho,$$

$$R_- = \left(\frac{1}{5} + \frac{5}{1} + \frac{4}{2}\right)\rho = 7.2\rho$$

$$R = \frac{R_+ + R_-}{2} = 8.6\rho.$$

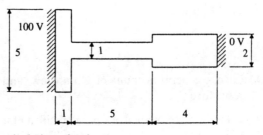

Fig. 1.15 The calculation of resistance.

1.8 A potential difference is applied, between a pair of vertical electrodes, to a rectangular conducting block of length 10 units, width 8 units, and depth 1 unit. Find the resistance per unit depth of the block. A nonconducting thin slot is made in the top surface half way between the electrodes to increase the value of its resistance (see Fig. 1.16). Estimate, using the *tubes-and-slices* method, the variation of the block's resistance with the slot penetration.

Solution
The simplest approximation will be as follows: (1) assume that the slices, all being vertical, give the same lower bound for the resistance

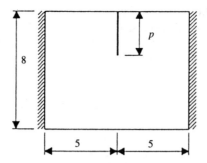

Fig. 1.16 A slotted block.

for all positions of the slot (see Fig. 1.17), (2) assume tubes with a horizontal current flow through the available cross section (see Fig. 1.18 – notice that current is excluded from the top section of the block). The results are summarized in Table 1.1 (all values are multiples of ρ and are in units of $\Omega \, m^{-1}$). The results obtained from the *finite-element* method are also included (in the bottom row of the table).

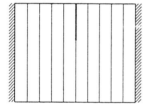

Fig. 1.17 Slices.

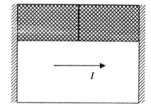

Fig. 1.18 Tubes.

Table 1.1

	p					
	0	1	2	3	4	5
R_+	$10/8 = 1.25$	$10/7 = 1.43$	$10/6 = 1.67$	$10/5 = 2$	$10/4 = 2.5$	$10/3 = 3.33$
R_-	1.25	1.25	1.25	1.25	1.25	1.25
R_{av}	1.25	1.34	1.46	1.625	1.875	2.29
R_{FE}	1.25	1.28	1.35	1.48	1.68	1.98

1.9 Alternative schemes for generating tubes and slices may be put forward. One such scheme, for a quadrilateral, is shown in Fig. 1.19. Think of the advantages and disadvantages of the proposed construction.

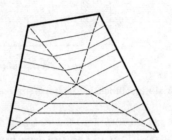

Fig. 1.19 The subdivision of a quadrilateral.

Electrostatic fields, capacitors   2

2.1 Energy storage

Capacitors are used in most electronic circuits as energy-storage devices. The amounts of energy are often small, but capacitors have the desirable property of rapid and almost loss-free energy exchange. They consist essentially of two conductors which are in close proximity but are insulated from each other. Fig. 2.1 shows a diagram representing a capacitor.

Fig. 2.1 A capacitor.

A potential difference, V, can be applied between the terminals. This will cause a transient current to flow, which 'charges' the capacitor. Positive and negative charges, Q, appear on the surfaces of the conductors close to each other, as shown in Fig. 2.2.

The amount of charge is proportional to the potential difference (p.d.), so that the ratio Q/V is independent of the supply and it is a property of the capacitor. We can write

(2.1)
$$\frac{Q}{V} = C$$

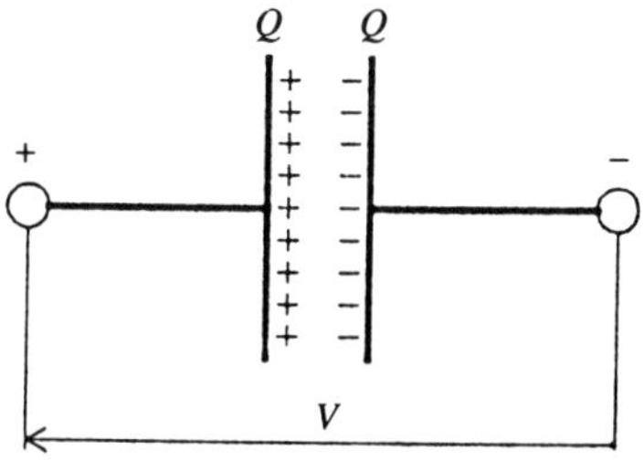

Fig. 2.2 A charged capacitor.

and call C the capacitance. The unit of charge is the coulomb, which is related to the unit of current by the relation 1 coulomb = 1 ampere second. The unit of capacitance is the farad, so that 1 coulomb/volt = 1 farad.

The stored energy is given by

(2.2)
$$U = \frac{1}{2} QV = \frac{1}{2} \frac{Q^2}{C} = \frac{1}{2} V^2 C.$$

The factor $\frac{1}{2}$ arises from the fact that Q and V are proportional to each other as shown in Fig. 2.3. The energy is given by the shaded area, since

(2.3)
$$U = \int_0^Q v \, dq = \int_0^Q \frac{q}{C} \, dq = \frac{1}{2} \frac{Q^2}{C}.$$

The energy is associated with the capacitor as a whole. The charges $+Q$ and $-Q$ exercise an attractive force on each other, so that a mechanical force has to restrain the conductors. One could, therefore, regard the energy as being associated with that restraining force. But in our discussion we shall regard the electrical action of the charges and the potential difference as being the 'cause' of the forces and of the energy. We can then enquire about the distribution of the electrical energy. It is reasonable to associate the energy with the charges and the space between them. We call this region the electric field, and much of this chapter will deal with the properties of such fields. Another way of looking at the energy distribution would be to think of the charges $+Q$ and $-Q$ as acting on each other at a distance. This would also involve the dimensions of the space between the charges, that is, the distance between them and the area of conductor on which there is charge, but it would not involve the space as a seat of energy.

There are, however, two reasons why the field view is preferable. The first relates to the design of capacitors. A similar design problem was met in Chapter 1, where knowledge was needed of the local distribution of the current and potential difference. The region was divided into tubes of current flow and slices of potential difference and a similar method may be used in electrostatics. The second reason will appear later in the book when electromagnetic waves carrying energy through apparently empty space

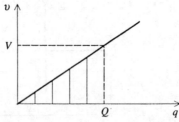

Fig. 2.3 The relationship between the charge, q, and the voltage, v.

are discussed. We then have to deal with energy in space whether we like it or not. It seems best therefore to become familiar with this difficult idea – even in electrostatics where this concept is not strictly necessary.

2.2 Electric flux

Chapter 1 dealt with the steady flow of a current in a homogeneous conductor. Physically, this flow is due to the motion of enormous numbers of electrons moving in an exceedingly complicated manner in all directions and at all sorts of velocities. But eqn (1.4) averaged the flow by a single, constant, drift velocity in the direction of the gradient of the potential difference. By considering a steady current, the electron motion was replaced by a homogeneous fluid of constant density. Such a fluid is incompressible. Suppose now that a source emits fluid at the rate of P m^3 s^{-1}. Let the source be surrounded by the surface of a wire mesh in the form of a sphere of radius r, as illustrated in Fig. 2.4, with the source at the centre of the sphere. By symmetry, the velocity is radially outwards, and it is given by

(2.4)
$$v = \frac{P}{4\pi r^2}.$$

Next, suppose that the wire mesh is not spherical but has an arbitrary shape, as in Fig. 2.5. The velocity at the mesh now has a tangential as well as

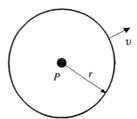

Fig. 2.4 A sphere surrounding a source.

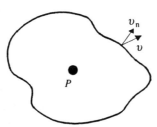

Fig. 2.5 An irregular wire mesh.

a normal component. Moreover, the magnitude of the velocity varies from place to place. Nevertheless, the total flow is still given by

(2.5)
$$P = \oint_S v_n \, \mathrm{d}S,$$

where v_n is the normal component of the velocity at the element $\mathrm{d}S$ of the surface. Equations (2.4) and (2.5) are interchangeable, and either can be derived from the other because they both relate to an incompressible fluid. Equation (2.5) is called Gauss's theorem and eqn (2.4) is called the inverse-square law. If there are no sources enclosed by the surface, or if there are as many sinks as sources, there will be no net outflow. Then eqn (2.5) becomes

(2.6)
$$\oint_S v_n \, \mathrm{d}S = 0.$$

For a steady current, therefore,

(2.7)
$$\oint_S J_n \, \mathrm{d}S = 0.$$

There is no net outflow or inflow of steady current through a closed surface.

We now come to the fundamental law of electrostatics, which relates the gradient of the potential difference to the electric charge. We have already noticed that for a capacitor the charge on the plates is proportional to the potential difference between the plates. This result is now generalized so that it can be applied to all electric-charge distributions, irrespective of their shape.

The unit of the gradient of potential difference is the volt per metre. Energy is given in units of volt coulomb (V C). Hence the units of volts per metre (V m^{-1}) can be written as energy/(coulomb metre). But energy/metre is a unit of force, so that the potential gradient is force/coulomb. Hence we can think of the gradient as the force on a 'probe' of unit charge. The fundamental law of electrostatics states that this force varies inversely as the square of the distance from a point charge.

As in the previous chapter, E is used for the negative gradient of the potential, that is, the force acts downhill like gravity. The law states that E is proportional to Q and

(2.8)
$$E = \frac{Q}{4\pi\varepsilon_0 r^2}.$$

It is illustrated by Fig. 2.6. $4\pi\varepsilon_0$ is a constant. The factor 4π is introduced because of the spherical geometry of Fig. 2.6, and ε_0 is a constant known as the permittivity of free space. Its dimensions are coulomb per volt per metre (C(V m)$^{-1}$) or farad per metre (F m^{-1}). In SI units the magnitude of ε_0 is

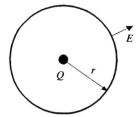

Fig. 2.6 An illustration of the inverse-square law.

8.854×10^{-12}. The magnitude depends on the use of the ampere as the unit of current. The coulomb is then defined as the ampere second (A s) and the volt as Joule per coulomb (J C^{-1}). We shall deal with the definition of the ampere when we consider the interaction of currents in a later chapter.

The inverse-square law is called Coulomb's Law after the great French experimentalist Charles Augustin de Coulomb who used a torsion balance to test the law in 1785. Unknown to Coulomb, the law had previously been verified by Henry Cavendish in a famous null experiment with a spherical capacitor. Cavendish found that when he electrified the outer sphere of the capacitor and connected it to the inner sphere no charge was transferred to the inner conductor. He calculated that he would have detected a charge if the inverse power in the law of force was outside the range 1.98–2.02. Modern experiments using a refined method similar to that used by Cavendish have narrowed the confidence limits to plus or minus one part in 10^9.

A comparison of eqns (2.8) and (2.4) shows that there is an analogy between the electrostatic field and the flow of an incompressible fluid, just as there is an analogy between current flow and an incompressible fluid. In current flow there were two field quantities: E, the potential gradient, and J, the current density, E being the force on the moving charges and J the flow per unit area. Equation (2.8) shows that in electrostatics E is the force per unit test charge, and the electric flux density, D, is now defined by the equation

(2.9)
$$D = \varepsilon_0 E.$$

This gives

(2.10)
$$D = \frac{Q}{4\pi r^2}$$

for the inverse-square law, and we can write

(2.11)
$$\Psi = 4\pi r^2 D = Q$$

for the flux crossing the spherical surface of Fig. 2.6. Since the flow is incompressible, the flux is independent of the shape of the surface and we can write

(2.12)
$$\oint D_n \, dS = Q.$$

Now consider the parallel-plate capacitor in Fig. 2.2, and assume for the present discussion that the charges are uniformly distributed on the plates. Then in the space between the plates

(2.13)
$$\oint D \, dS = DS = Q$$

and

(2.14)
$$\int E \, dl = El = V,$$

where l is the distance between the plates. Then the capacitance is given by

(2.15)
$$C = \frac{Q}{V} = \frac{DS}{El} = \frac{\varepsilon S}{l},$$

and the energy is given by

(2.16)
$$U = \frac{1}{2} QV = \frac{1}{2} DSEl = \frac{1}{2} \varepsilon E^2 Sl = \frac{1}{2} \frac{D^2}{\varepsilon} Sl,$$

where Sl is the volume occupied by the field, so that this equation relates the energy to the region in which there is a field. Later in this chapter we shall relax the condition of uniform charge density on the plates and shall show how the capacitance is calculated, including an edge effect which is associated with a fringing flux.

2.3 Tubes and slices in the electrostatic field

Equations (2.13) and (2.14) apply to a parallel-plate capacitor without edge effects. The field has a very simple distribution, and it is in the same direction everywhere, and it is of constant strength. The field can be divided into tubes of flux and slices of potential difference as shown in Figs 2.7 and 2.8.

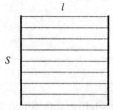

Fig. 2.7 Tubes of flux.

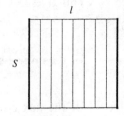

Fig. 2.8 Slices of potential.

The tubes could be bounded by thin sleeves which are impervious to flux, and the slices could be bounded by thin equipotential surfaces. Such boundaries would not disturb the flux or the potential gradient. The capacitance of the tubes could be obtained by addition, because the tubes all have the same potential difference, V. So

(2.17)
$$C = \frac{Q}{V} = \sum \frac{q_t}{V} = \sum C_t,$$

where q_t is the charge at the ends of a tube.

The capacitance of the slices could be obtained by addition of the inverse capacitances of the slices, which all have the same charge, Q

(2.18)
$$\frac{1}{C} = \frac{V}{Q} = \sum \frac{v_s}{Q} = \sum \frac{1}{C_s},$$

where v_s is the potential difference across a slice.

In this example, nothing is gained by the subdivision, but in general the field distribution will be more complicated. The subdivision then becomes very useful.

2.4 Calculation of capacitance using tubes and slices

Every tube and every slice has a potential difference between its two equipotential surfaces and an associated pair of charges on these surfaces. The product of potential difference and charge can associate a positive energy with the tubes (and with the slices), and the sum of these energies is the system energy. It is useful to describe this system energy in terms of the system (or circuit) parameter of capacitance. Consider a piece of tube of length δl and average cross section, δS. The capacitance is

(2.19)
$$\frac{\delta Q}{\delta V} = \frac{D \delta S}{E \delta l} = \varepsilon \frac{\delta S}{\delta l}.$$

If there are m such pieces along a tube, these capacitances are in series and are given by

(2.20)
$$\frac{1}{C} = \sum_1^m \frac{\delta l}{\varepsilon \delta S}.$$

If there are n tubes in parallel the total capacitance is

(2.21)
$$C = \sum_1^n \left(\sum_1^m \frac{\delta l}{\varepsilon \delta S} \right)^{-1}.$$

If instead we work with the slices, for n slices in parallel,

$$(2.22) \qquad C = \sum_1^n \frac{\varepsilon \delta S}{\delta l},$$

and for m slices in series

$$(2.23) \qquad C = \left[\sum_1^m \left(\sum_1^n \frac{\varepsilon \delta S}{\delta l} \right)^{-1} \right]^{-1}.$$

If all the tubes and slices are orthogonal, and if m and n are large, then these two values of capacitance will be nearly equal. If the sums were replaced by integrals, they would be equal, but in a numerical method m and n are necessarily finite.

There is a striking similarity between eqns (2.20)–(2.23) used for calculating capacitance and eqns (1.11)–(1.15) in Chapter 1 used for calculating resistance. We can see that the analogue of current is electric flux and conductance is equivalent to capacitance, where conductance is the reciprocal of resistance. The analogue of the conductivity, σ, is the permittivity, ε. The subdivision into tubes produces the lower bound, whereas slices give the upper bound of capacitance.

Figure 2.9 illustrates the cross section of a tubular capacitor. Owing to symmetry, only a quarter of the system needs to be investigated, as shown in Fig. 2.10, and the total capacitance will then be four times the value calculated for Fig. 2.10, since four identical capacitors are connected in parallel.

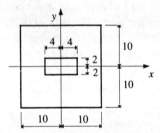

Fig. 2.9 A tubular capacitor.

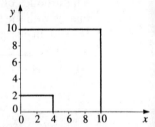

Fig. 2.10 One quadrant.

By analogy with the examples discussed in Chapter 1 (for example, Fig. 1.13) an approximate distribution of tubes and slices can be suggested, based on the system of quadrilaterals shown in Fig 2.11. By following the technique described in Section 1.5, the two distributions illustrated in Figs 2.12 (tubes) and 2.13 (slices) can be arrived at. The combined system of tubes/slices is shown in Fig. 2.14, this gives values for the upper and lower bounds of $1.5386\varepsilon_0$ and $1.2303\varepsilon_0$, respectively. Note that all results are

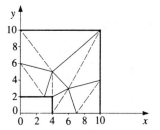

Fig. 2.11 A system of quadrilaterals.

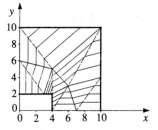

Fig. 2.12 A distribution of tubes.

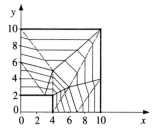

Fig. 2.13 A distribution of slices.

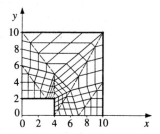

Fig. 2.14 Tubes and slices.

given per unit depth. The average capacitance is $1.3845\varepsilon_0$ and the error band is ± 11.1 per cent. Figure 2.15 shows an improved subdivision which gives bounds of $1.4656\varepsilon_0$ and $1.3252\varepsilon_0$, with an average value of $1.3954\varepsilon_0$ and an error band ± 5.0 per cent. Notice that the improved distribution exhibits an improved orthogonality. The average value is hardly changed, but the error band is narrower, and this gives additional confidence to the designer. Finally, the field plot obtained from a finite-element program is shown in Fig. 2.16. The capacitance calculated from this solution is $1.3979\varepsilon_0 \pm 0.32$ per cent.

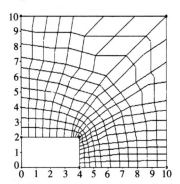

Fig. 2.15 An improved distribution of tubes and slices.

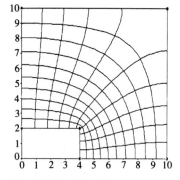

Fig. 2.16 A finite-element solution for Fig. 2.10.

2.5 Parallel-plate capacitors with fringing

In Sections 2.2 and 2.3 we looked at a parallel-plate capacitor and we found a very simple equation for calculating its capacitance. For this calculation, we assumed that the field between the plates is uniform. This is, however, not true at the edges and Fig. 2.17 shows a more accurate plot of the field near the edge of a parallel-plate capacitor. The field plot of Fig. 2.17 represents an almost correct distribution of tubes and slices since they are nearly orthogonal everywhere and the error band is thus very narrow (less than 1 per cent). In practice, such a highly accurate solution may not be necessary, and simpler and faster calculations may be performed following the techniques put forward in Sections 1.5 and 2.4. Such exercises are left to the reader.

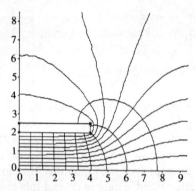

Fig. 2.17 A parallel-plate capacitor with fringing.

2.6 System energy

The division into tubes gives a lower bound, C_-, for the total capacitance, and the division into slices gives an upper bound, C_+. This is analogous to the resistance problem in Chapter 1. Unless the tubes are drawn strictly along the flow they will disturb the flux by lengthening its path and therefore they will reduce the capacitance just as the tubes increase the resistance (reduce the conductance) to current flow. Similarly, the slices disturb the voltage distribution and this increases the capacitance because it shortens the flux path.

Subdivision into tubes and slices enables the statement about the distribution of energy contained in eqn (2.16) to be generalized. A tube is shown in Fig. 2.18. Notice first of all that the tube has equal magnitudes of charge at its ends because no flux crosses the curved surface bounding the tube. The flux is equal to the charge at its ends. Let the tube now be

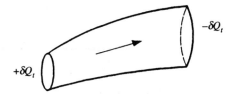

Fig. 2.18 A tube.

subdivided into a succession of thin slices as shown in Fig. 2.19. Each slice is a parallel-plate capacitor of cross section δS and length δl. Because of the fine subdivision both D and E can be regarded as constant. Hence

(2.24)
$$\delta U = \frac{1}{2} \delta Q \, \delta V = \frac{1}{2} D \, \delta S E \, \delta l = \frac{1}{2} \varepsilon E^2 \, \delta S \, \delta l = \frac{1}{2} \frac{D^2}{\varepsilon} \, \delta S \, \delta l.$$

Hence it can be asserted that the energy density per unit volume is given by $\frac{1}{2}\varepsilon E^2$ or $\frac{1}{2}D^2/\varepsilon$. This statement is subject to the volume being defined by δS and δl in accordance with the tube–slice subdivision.

The energy is associated with a force. Consider the small section of tube shown in Fig. 2.19. The field, E, is from left to right and it pushes the charge $+\delta Q$ to the right and the charge $-\delta Q$ to the left. The tube is therefore in a state of tension. The tensile force is $\frac{1}{2}E\,\delta Q$ and the tensile stress is therefore $\frac{1}{2}E\,\delta Q/\delta S = \frac{1}{2}ED$ N m^{-2}. The reason for the factor $\frac{1}{2}$ is interesting. Since E is force per unit charge the force might be expected to be $E\,\delta Q = ED$, but this would have been wrong, because E is the field in the region between $-\delta Q$ and $+\delta Q$. To calculate the tensile force E is needed at the place where the charges are. Note that the field *between* the plates of a capacitor is associated with *both* the positive and the negative charges. The force on each charge is due to the other charge and that is $\frac{1}{2}E$ per unit charge. The same result is obtained by averaging the field at the surface of a conductor, as illustrated in Fig. 2.20. Inside the conductor, there is no field once any transient current has died away. Outisde the conductor, the field is E. Hence the average field at the surface is $\frac{1}{2}(E+0) = \frac{1}{2}E$. If the surface density of charge is given by σ (C m^{-2}) the outside flux density will be

(2.25)
$$D = \sigma,$$

Fig. 2.19 A tube subdivision.

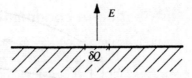

Fig. 2.20 A conductor surface.

and the force trying to pull the charge out of the surface will be due to a tensile stress

$$\tfrac{1}{2}E\sigma = \tfrac{1}{2}ED.$$ (2.26)

In electrostatics there can be no tangential field at the surface of a conductor, because such a field would cause a current to flow. This means that the surface charges on a conductor must arrange themselves so as to cancel all tangential forces. Figure 2.20 is of course incomplete in showing only a part of a conductor. Moreover, the flux emerging from the surface of the conductor must end on another conductor carrying a charge of the opposite sign. A totally isolated charge is an impossibility. Moreover it is impossible for a flux tube to start and end on the same conductor, because this would imply a potential difference between the two parts of the conductor. Such a potential difference would cause an electric current to flow, and this would no longer be an electrostatic problem.

Conversely, note that in problems of current flow, like those in Chapter 1, where there is a potential difference between parts of the same conductor, there will in general be flux tubes, as illustrated in Fig. 2.21. This means that there will be positive and negative surface charge on the conductor which will therefore exhibit capacitance as well as conductance. However, the amount of capacitance will be small because the potential difference along the conductors will be small and the surface charge will be small.

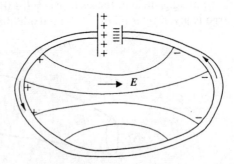

Fig. 2.21 The potential difference and flux tubes.

2.7 The charge distribution in conductors

So far, it has been assumed that the charge distribution is always on the surface of conductors. This is reasonable because any distribution of net positive or negative charge in a conductor will exert inverse-square-law forces of repulsion causing the charges to separate as far as possible. This means that they will move to the outer surface. However, since capacitors are often used in high-frequency applications, it is important to investigate the time taken by the charges to move to the surface.

Consider a volume charge distribution, as illustrated in Fig. 2.22. For simplicity, let there be a negative charge, Q, uniformly distributed in a spherical region of radius r. At the surface of the region there will be a radially inward field

$$E = \frac{Q}{4\pi\varepsilon_0 r^2}.$$

This will cause a conduction current of density

$$J = \sigma E,$$

where σ is the conductivity. The incoming current will decrease the negative charge in the region at the rate

(2.27)
$$-\frac{dQ}{dt} = I = 4\pi r^2 J = \frac{\sigma Q}{\varepsilon_0}.$$

Hence

(2.28)
$$Q = Q_0 \exp(-\sigma t/\varepsilon_0),$$

where Q_0 is the initial charge. The time constant of the exponential decay is ε_0/σ. Taking σ for copper as 10^8 gives a time constant of the order of 10^{-19} s. This is so enormously fast that it is safe to conclude that there cannot be any volume distribution of charge in copper or in any conducting material. Hence tubes of electric flux cannot start inside a conductor. Of course this does not mean that there is no electric flux in conductors when they are carrying current, but the tubes of flux must be closed on themselves

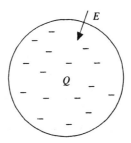

Fig. 2.22 A volume charge.

or be closed at the surface of the conductors. Tubes of flux closed on themselves occur in time-varying electric fields which are discussed in Chapters 5 and 6.

2.8 Insulators

In discussing capacitors the material between the plates has so far been ignored. In fact, it has been assumed that there is no such material or that it has no effect electrically. The constant ε_0 which determines the capacitance and the energy storage is an inherent property of space, or more accurately of space and time. Now consider the effect of an insulating material between the plates. All materials contain both positive and negative electric charge. Generally, the negative electrons are symmetrically disposed in shells around the positive nucleus of the atoms as shown in Fig. 2.23, so that they have no external electric field. However such atoms can be polarized as indicated in Fig. 2.24. There are also some materials, notably water, in which the molecules are inherently polarized. In the absence of an electric field, the random positions of the molecules cancel their overall electrical effect, but when an external field is applied there is an alignment of molecules which reinforces the field. Moreover the polarization of the individual molecules is increased. The bulk effect is illustrated in Fig. 2.25.

The difference between the movement of charge in insulators and in conductors is that in insulators the charges are bound to the material, whereas in conductors the charge is free to travel through the material. It is therefore useful to distinguish between *bound* and *free* charge. The bulk effect outside the material is the same, because in both cases there is a

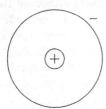

Fig. 2.23 An atom.

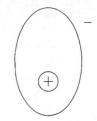

Fig. 2.24 A polarized atom.

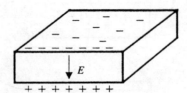

Fig. 2.25 Bulk polarization.

surface charge. Inside the material there is a difference because there is no electrostatic field inside a conductor, whereas an insulator requires an internal field to maintain its polarization.

2.9 The effects of polarization

Now consider the insertion of a polarizable or dielectric material between the plates of a capacitor as shown in Fig. 2.26. The free charge on the conducting plates is $\pm Q$, and the bound charge on the surfaces of the dielectric is $\pm q$. The field, E, in the dielectric material is given by

(2.29)
$$E = \frac{Q-q}{\varepsilon_0 S}.$$

If the insulator fills the entire space between the plates, the potential difference is given by

(2.30)
$$V = \int -E \, dl = \frac{Q-q}{\varepsilon_0 S} l.$$

The capacitance is given by

(2.31)
$$C = \frac{Q}{V} = \frac{Q}{Q-q} \frac{\varepsilon_0 S}{l}.$$

The induced charge, q, has increased the capacitance by the ratio $Q/(Q-q)$. The charge q has been produced by the internal field E and this field is proportional to $Q-q$ as seen in eqn (2.29). The dimensionless ratio $q/(Q-q)$ is a property of the polarizability of the material and it is called the electric susceptibility, χ_e. Thus, for a parallel-plate capacitor,

(2.32)
$$\chi_e = \frac{q}{Q-q}.$$

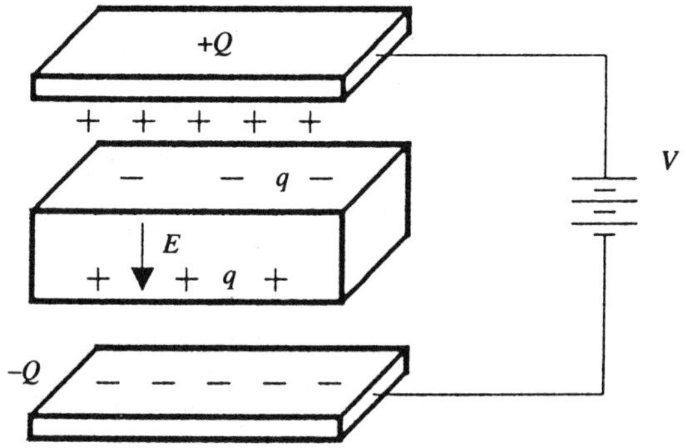

Fig. 2.26 A capacitor with a dielectric.

The increase in capacitance in eqn (2.31) is given by

(2.33)
$$\frac{Q}{Q-q} = 1 + \frac{q}{Q-q} = 1 + \chi_e = \varepsilon_r,$$

where ε_r is called the relative permittivity. The capacitance can be written

(2.34)
$$C = \frac{\varepsilon_0 \varepsilon_r S}{l} = \frac{\varepsilon S}{l},$$

where ε is called the *permittivity* and it is the product of the *permittivity of free space* and the *relative permittivity*. It should be noted that ε_r is a dimensionless ratio and is just a number, whereas ε_0 and therefore ε have units of farad per metre (F m^{-1}), so that their numerical magnitude depends on the choice of the basic units.

2.10 Partial and total fields

A word of explanation is needed about the field, E, inside the polarizable material. We are describing bulk effects and our theory cannot deal with interatomic distances. To be *inside* the material means therefore to be in a cavity of the material.

Consider a cavity in the form of a cylinder with its axis along the field as in Fig. 2.27. The polarization will produce surface layers as indicated. Suppose the cylinder is long and narrow ($d/l \ll 1$), then the field in the centre of the cylinder will be virtually unaffected by the charge at the top and the bottom of the cylinder. This is the same type of field, E, as is shown in Fig. 2.26. Suppose, on the other hand, that the cylinder is short and wide ($d/l \gg 1$), then the field will be strongly affected by the charges at the ends of the cylinder. In fact these charges will cancel the induced charges on the surfaces of the insulating slab. The field in such a cylinder will be the same as if there were no polarizable material and it will be $\varepsilon_r E$. These results are

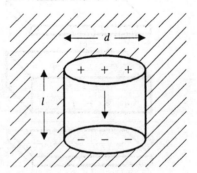

Fig. 2.27 A cylindrical cavity.

illustrated in Fig. 2.28. Notice that it is the shape of the cylinder that matters, and not its size.

To distinguish between the two cases the notion of tubes of flux is used. Define the flux density by

(2.35)
$$D = \varepsilon_0 \varepsilon_r E.$$

This implies that tubes of D terminate on conductor surfaces, but they pass right through polarizable material. The sources of D are free charges on conductors. On the other hand, E is affected by the bound charges as well as by the free charges. Hence, D is the partial field of free charges, and E is the total field of free and bound charges.

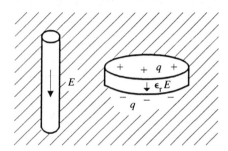

Fig. 2.28 Two types of cavity.

2.11 Boundary conditions

Now consider the field near the surface of a dielectric of relative permittivity ε_r, as illustrated in Fig. 2.29. The normal component of D is continuous, because no tubes of D can terminate at the layer of bound charge. Hence

(2.36)
$$D_{n1} = D_{n2},$$

(2.37)
$$E_{n1} = \varepsilon_r E_{n2}.$$

The tangential component of E is continuous, because the bound charges at

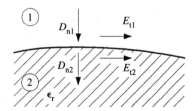

Fig. 2.29 The surface of a dielectric.

the surface affect the normal component of E but not the tangential component, as illustrated by Fig. 2.30. Hence

(2.38)
$$E_{t1} = E_{t2}.$$

Combining eqns (2.37) and (2.38)

(2.39)
$$\frac{E_{t1}}{E_{n1}} = \frac{1}{\varepsilon_r} \frac{E_{t2}}{E_{n2}}.$$

This is illustrated in Fig. 2.31. Hence,

(2.40)
$$\tan \theta_2 = \varepsilon_r \tan \theta_1,$$

where θ_1 and θ_2 are the angles to the normal of the surface.

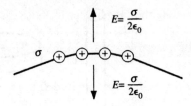

Fig. 2.30 Surface charges.

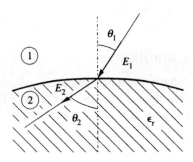

Fig. 2.31 Boundary conditions.

Exercises

2.1 Define the term capacitance and derive a simple formula for calculating the capacitance of a parallel-plate capacitor with edge effects ignored. If a polarizable material is inserted between the plates how will it affect the capacitance?

2.2 Four capacitors are connected as shown in Fig. 2.32. Find the effective capacitance, in terms of C, between the terminals A and B.
(*Answer 2C.*)

Fig. 2.32

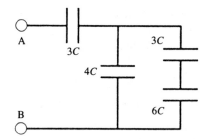

2.3 A capacitor consists of a pair of plane parallel metal plates each 100×100 mm^2 in area and separated by a distance of 1 mm. A sheet of dielectric of thickness (a) 1 mm or (b) 0.5 mm, and with a relative permittivity of 5 is inserted between the plates. Neglecting edge effects, estimate the capacitance for both cases. How would the capacitances change if all the dimensions were reduced by a factor of ten? (*Answer* (a) 443 pF, (b) 148 pF. They would be reduced by a factor of ten.)

2.4 A silicon chip of resistivity 1 Ω m has dimensions $200 \times 10 \times 10$ μm^3. A potential difference of 10 V is applied over the longest dimension of the chip. Calculate the power dissipated. (*Answer* 50 μW.)

2.5 A cylindrical capacitor is illustrated in Fig. 2.33. Neglecting edge effects, derive an expression for the capacitance.

Solution

$$D(2\pi rl) = Q, \quad E = \frac{Q}{2\pi\varepsilon_0 rl},$$

$$V = \int_a^b -E\,\mathrm{d}r = \frac{Q}{2\pi\varepsilon_0 l}\ln(a/b),$$

$$C = \frac{Q}{V} = \frac{2\pi\varepsilon_0 l}{\ln(a/b)}.$$

Fig. 2.33 A cylindrical capacitor.

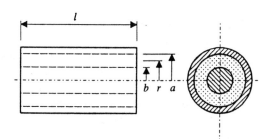

2.6 In a spherical capacitor the radius of the inner conductor is b, and the inner radius of the outer conductor is a. The space between the conductors is filled with a material of permittivity ε_r. Calculate the capacitance.

(*Answer* $4\pi\varepsilon_r\varepsilon_0/(1/b-1/a)$.)

2.7 Derive expressions for the energy stored in a capacitor.

2.8 Explain carefully what is meant by a tube of electric flux. What is the connection between this concept and the inverse-square law of force between charges?

2.9 Discuss the use of *tubes and slices* in calculations of the dual bounds of a capacitance. Discuss the analogy between such calculations and estimations of resistance using the same technique. Demonstrate how the calculations lead to series/parallel connections of simple parallel-plate capacitors. What are the computational advantages of the method?

2.10 A tubular capacitor is illustrated in Fig. 2.34. Estimate the capacitance using the tubes-and-slices method, and compare this result with the value obtained using a finite-element program, $C = 14.24\varepsilon_0$ F m^{-1}.

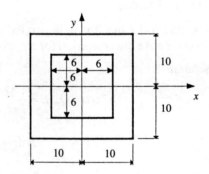

Fig. 2.34 A tubular capacitor.

Solution

Because of symmetry only one eighth of the system needs to be investigated, and the total capacitance will thus be eight times the value calculated for one segment. Some simple assumptions regarding the shape of the electric flux (tubes) and the distribution of equipotential lines (slices) can be made, so that the appropriate capacitances can be calculated by hand. A possible procedure is shown graphically in Fig. 2.35. The capacitance can now be calculated as follows

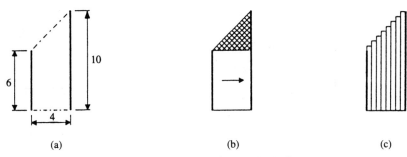

Fig. 2.35 (a) A capacitance segment, (b) tubes, and (c) slices.

$$C_- = \tfrac{6}{4}\varepsilon_0 = 1.5\varepsilon_0 \text{ F m}^{-1},$$

$$C_+ = \left(\sum \frac{1}{C_s}\right)^{-1}$$

$$= \left[\frac{0.5}{\varepsilon_0}\left(\frac{1}{6.5} + \frac{1}{7} + \frac{1}{7.5} + \frac{1}{8} + \frac{1}{8.5} + \frac{1}{9} + \frac{1}{9.5} + \frac{1}{10}\right)\right]^{-1}$$

$$= 2.02\varepsilon_0 \text{ F m}^{-1},$$

$$C_{av} = \frac{C_- + C_+}{2} = \frac{1.5 + 2.02}{2}\varepsilon_0 = 1.76\varepsilon_0 \text{ F m}^{-1},$$

and finally $C = 8 \times 1.76\varepsilon_0 = 14.08\varepsilon_0$ F m^{-1}.

2.11 Fig. 2.17 shows the field at the edge of a pair of long parallel plates. If the width of the plates is large compared with the distance between them, the field between the plates is practically uniform and outside the plates it is negligible. It is then reasonable, when finding the capacitance, to use a simple equation which ignores edge effects (eqn 2.15). Demonstrate that by doing so a lower bound is used for the capacitance. (*Hint* Think of the distribution of tubes which is implied by neglecting the edge effects.)

Using the method of tubes and slices, estimate the increase of capacitance caused by the fringing field at the edge of the plates as a function of the ratio of the plate width to the separation distance.

Hint find the capacitance per unit depth for the capacitor shown in Fig. 2.36. Due to symmetry, only a quarter of the system needs to be investigated, as illustrated above. Notice how symmetry introduces series/parallel connections of the component capacitors, which are identical, and thus simplifies the calculations. It is also interesting to note that this is now an *open-boundary* problem because there is no

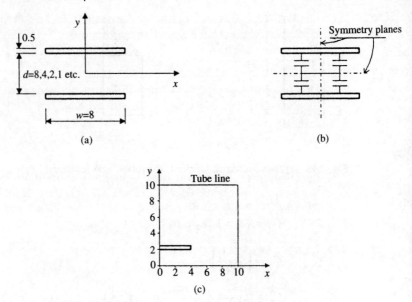

Fig. 2.36 (a) A parallel-plate capacitor, (b) the symmetry conditions, and (c) a computational model.

physical boundary to contain the electric flux. However, the field will decay very quickly away from the capacitor and it is justifiable to assume that the field can be enclosed in a 10×10 box formed by a tube line so that no flux escapes out of the box. The effect of changing the size and/or position of the box on the accuracy of solution can be studied separately, but this is not part of this exercise. Calculations can be performed using the tubes-and-slices techniques studied in previous examples in Chapters 1 and 2. Typical results are summarized in Table 2.1.

Table 2.1

	w/d				
	1	2	4	8	16
Capacitance[a] (using eqn (2.15)) (edge effect ignored)	1	2	4	8	16
Capacitance[a] (edge effect included)	1.83	3.14	5.42	9.77	18.1
Increase in capacitance (%)	83	57	35.5	22.1	13.1

[a] All capacitances are multiples of ε and are per unit depth (F m^{-1}).

2.12 Discuss the behaviour of charges in insulators and in conductors. Hence define *bound* and *free* charges. Demonstrate that there may be no electric field inside a conductor.

2.13 A long conducting cylinder of radius R carries a charge Q ($C\,m^{-1}$). Determine the electric field strength in the following cases (a) inside the cylinder, (b) at the surface of the cylinder, (c) just outside the cylinder, and (d) at a distance r from the cylinder.
(*Answer* (a) 0, (b) $Q/4\pi\varepsilon_0 R$, (c) $Q/2\pi\varepsilon_0 R$, (d) $Q/2\pi\varepsilon_0 r$.)

2.14 Two small charged spheres carrying charges Q_A and Q_B are separated by a distance of 10 cm; the force between them is 10 N. The distance between the spheres is reduced to 5 cm. What is the new force between the charged spheres?
(*Answer* 40 N.)

2.15 Explain why D is the partial field of free charges and E is the total field of free and bound charges.

2.16 Define the terms *permittivity*, *permittivity of free space*, and *relative permittivity*.

2.17 Discuss the boundary conditions at a dielectric surface. Estimate the total electric field, E, inside a thin plate of relative permittivity, ε_r, inserted across a uniform field, E_0.
(*Answer* E_0/ε_r.)

3.1 Electrokinetic energy

An idea which unifies all the topics in this book is the idea of energy and its distribution in space and time. In Chapter 1 there was a discussion of the conversion of electrical energy into heat in terms of the property of resistance. This conversion process is irreversible because the steady average motion of the free charges in the conductor is converted into the random vibrations of the conductor lattice. A mechanical analogy of the process is the diffusion of a viscous liquid through a porous material. The rate of energy conversion is the product of the current and the potential difference between the entry and exit ports of the current. The conversion process occurs inside the conducting material and it can be analysed in terms of tubes of current and slices of potential difference.

In Chapter 2 there was a discussion of the potential energy associated with the position of electric charges, which gives rise to the property of capacitance exhibited by capacitors. This energy is stored in capacitors when a potential difference separates positive and negative charges on the plates of the capacitor and it can be recovered when the capacitor is discharged. The mechanical analogy is the stretching of a spring; but, whereas elastic spring energy is associated with the material of the spring, the capacitance energy is associated with the space between the charges which need not contain any material substance, although it does contain energy. The inverse-square law of force between charges suggests that the energy space is filled by an electric flux associated with the potential difference, so that the energy distribution can be divided into tubes of flux and slices of potential difference.

The potential difference drives both the steady current and the electric flux. This implies that during the charging or discharging of a capacitor there will be some ohmic loss. The capacitor will exhibit resistance as well as capacitance; but, as mentioned in Chapter 2, the loss of energy is extremely small compared with the amount of energy stored, so that for many calculations a physical object called a capacitor can be represented solely by the storage called capacitance. Similarly a resistor will have some surface charge associated with the potential difference. Hence it will have some capacitance, but the energy associated with this will generally be negligible. Hence such a resistor can often be represented by a 'pure' resistance.

This is often the case, but not invariably – because a third kind of energy must now be introduced. If two conductors carrying steady currents are placed parallel to each other, there is a force between them, as pictured in Fig. 3.1. Figure 3.1 is of course incomplete because steady currents must flow in complete loops; but in Fig. 3.1(a) the ends of the loop could be some distance away, and in Fig. 3.1(b) there could be a return path, carrying a current $2I$, a considerable distance away. Notice that like currents attract and unlike currents repel each other. This is the opposite tendency to that observed by like charges, which repel, and by unlike charges which attract each other. Experiments show that the forces depend on the current, but not on the potential difference. In fact, the unit of current, the ampere, is defined in terms of the force between parallel currents. This force is given by

(3.1)
$$F = \frac{\mu_0 I_1 I_2}{2\pi a} \quad (\text{N m}^{-1})$$

where I_1 and I_2 are currents in arbitrary units, μ_0 is a dimensional constant, a is the distance between the currents and the factor 2π arises because long linear currents are associated with a cylindrical geometry. In SI units the ampere is defined as 'that constant current, which if maintained in two straight parallel conductors of infinite length, of negligible circular cross-section, and placed 1 metre apart in vacuum, would produce between these conductors a force equal to 2×10^{-7} newton per metre length'. Inserting these values into eqn (3.1)

(3.2)
$$2 \times 10^{-7} = \frac{\mu_0}{2\pi} \frac{1 \times 1}{1}$$

or

(3.3)
$$\mu_0 = 4\pi \times 10^{-7}.$$

The 'dimensions' of μ_0 are

(3.4)
$$\frac{\text{force}}{\text{metre}} \frac{\text{metre}}{(\text{current})^2} = \frac{\text{energy}}{\text{metre}(\text{current})^2}.$$

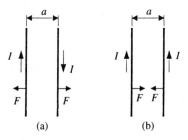

(a)   (b)

Fig. 3.1 Two long parallel currents.

In SI units

$$\frac{\text{energy}}{(\text{current})^2} = \frac{\text{joules}}{(\text{ampere})^2} = \text{henry}.$$

The unit of μ_0 is therefore henry per metre (H m^{-1}). The henry defines the energy per (unit current)^2 and is a unit of *inductance*.

Since μ_0 depends on the current and ε_0 depends on the charge, there must be some connection between them. From the inverse-square law the 'dimensions' of ε_0 are

(3.5)
$$\varepsilon_0 = \frac{(\text{charge})^2}{\text{force}(\text{metre})^2} = \frac{(\text{ampere second})^2}{\text{energy metre}}.$$

Hence the 'dimensions' of the product $\varepsilon_0 \mu_0$ are

(3.6)
$$\varepsilon_0 \mu_0 = \left(\frac{\text{second}}{\text{metre}}\right)^2 = \frac{1}{(\text{velocity})^2}.$$

Since $\mu_0 = 4\pi \times 10^{-7}$ defines the ampere and therefore the coulomb, a measurement of force between known charges will provide the numerical value of ε_0. This has already been stated as $\varepsilon_0 \approx 8.854 \times 10^{-12} \text{ F m}^{-1}$. Hence the velocity in eqn (3.6) is given by $v \approx 10^8 \text{ m s}^{-1}$. This is the *velocity of light*, as derived from optical observations which make no explicit use of electrical quantities. Clearly there is something of immense significance in this relation between electrostatic forces and the forces between currents.

But we shall proceed cautiously and first explore the nature of the force between currents. Notice that there is a certain similarity between the forces between currents and the forces between charges. Both forces depend on geometrical factors such as the position of the conductors and the distance between them. Also these forces can act in a vacuum, and they are therefore not confined to material substances. Secondly, notice that the direction of the force between currents is such as to increase the mutual energy between them. Currents flowing in the same direction will reinforce each other and the force tries to bring them together. Unlike currents will cancel each other, and the force drives them apart. This is the opposite kind of action to that which occurs with electric charges – where like charges repel, and unlike charges attract. It has been seen that the energy of such charges is the potential energy associated with the potential difference between them and the analogy of a stretched spring was used earlier. The energy of the current is not a potential energy but it is a kinetic energy because it is associated with moving charges in the form of steady currents. Now in mechanics, the forces due to kinetic energy act to increase the energy, whereas the forces due to potential energy act to decrease that energy. Fundamentally, this is a consequence of d'Alembert's principle which states that an accelerating mass is equivalent to a negative external force. Newton's law,

(3.7)
$$F = ma,$$

was written by d'Alembert as

(3.8)
$$F - ma = F - F' = 0.$$

Here, F is an external force due to the potential energy, and F' is an equivalent force due to the kinetic energy. The comparison of the forces associated with electric currents and the forces on moving masses in mechanics suggests that there is a kinetic energy associated with currents. However, in mechanical motion, the energy is inherent in the moving masses, whereas in currents the energy is a system parameter distributed in space. Thirdly, notice that both the potential and the kinetic energy are associated with forces and therefore they can be converted into each other. In mechanics a mass can be suspended from a spring. Such a system will oscillate, and during its motion there will be a continuous interchange between the elastic energy of the spring and the kinetic energy of the mass. Circuits consisting of capacitance and inductance show an analogous behaviour.

3.2 Electricity and magnetism

The SI definition of the ampere describes a highly idealized experiment, which involves infinitely long conductors of negligible cross section. This is reminiscent of the inverse-square law of force between point charges of negligible volume. In the language made familiar by Einstein these are 'thought experiments', which are to be tested by their consequences. For example, the inverse-square law can be tested by observing that there is no electric field inside a charged hollow spherical conductor. The SI definition of current is based on experiments undertaken in 1820 by André Marie Ampère in Paris. Ampère was interested in the relationship between magnetism and electricity, an interest which had been aroused by the discovery in the same year by the Danish physicist Hans Christian Oersted that there was a force on a magnetic compass needle in the vicinity of an electric current. Until that discovery electricity and magnetism had been studied as independent subjects.

Magnetism had been based on the effects of naturally occurring 'permanent' magnets. A very thorough experimental and theoretical investigation of the behaviour of such magnets was published as early as 1600 by an English medical doctor called William Gilbert. He drew particular attention to the external field around magnets, which he explored by means of a small magnetic needle, and which he described in terms of curved lines of force converging on, or diverging from, the two opposite poles of a magnet. He experimented chiefly with spherical magnets and inferred that the earth was itself a large magnet. Gilbert's field theory did not gain wide acceptance because of the influence of Newton's subsequent theory that gravitational forces were inherent in material masses which acted on each other, at a distance, with an inverse-square law of force. We have already noted that Cavendish and Coulomb used the same type of law to explain electrostatic effects. Coulomb also found, by use of his torsion balance, that magnetic forces could be similarly attributed to the action at a distance between magnetic poles and that the inverse-square law again applied.

The three phenomena of gravitation, electricity, and magnetism are not identical. Gravitational forces are always forces of attraction, whereas electrical forces exhibit repulsion as well as attraction. Magnetic forces are similar to electrical ones in this respect, nevertheless, there are no magnetic conductors and it is impossible to separate the two types of polarity, which always occurr in pairs. The inverse-square law, therefore, did not provide a physical explanation, but it did provide a common mathematical system for the three phenomena, as long as these were thought to be independent.

Oersted's discovery changed all this. Ampère deduced that magnetic energy is the kinetic energy of moving charges. The reason why poles occur in pairs is that fundamentally magnetic sources are due to current loops. Ampère postulated that there were 'molecular' currents inside magnetic materials. This was a remarkable guess, which was substantiated a century later in terms of electronic orbits and the spin motion of electrons.

Ampère's experimental results can be summarized by his equivalence theorem which states that the external effects of small magnetic dipoles and small loops of constant current are identical. The external effects must be observed at a distance which is large compared to the dimensions of the dipole and the loop, because the internal structure of dipoles would be modelled differently from that of current loops. This equivalence is illustrated in Fig. 3.2. The current I has a negligible cross section, but it

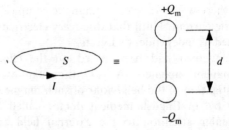

Fig. 3.2 The equivalence of a current loop and a dipole.

forms a loop of area S. The magnetic dipole has pole strengths of $\pm Q_m$ separated by the distance d. The *dipole moment* is $Q_m d$. Its direction is given by d, which is perpendicular to the area S. The relationship of the positive to negative pole strength, that is, the direction of the dipole moment, is related to the direction of the current by the motion of a right-handed screw.

In order to fix the dimensions and unit size of the pole strength we write

(3.9)
$$\mu_0 I S = Q_m d.$$

This gives

(3.10)
$$Q_m = \frac{\text{joules}}{(\text{ampere})^2} \frac{\text{ampere}}{\text{metre}} \frac{(\text{meter})^2}{\text{metre}} = \text{volt second}.$$

The volt second is named the weber. Equation (3.9) introduces a remarkable symmetry between the units of electrostatics and magnetostatics. The electric charge is the *coulomb* or *ampere second* and the magnetic 'charge' is the *weber* or *volt second*. Moreover this symmetry is also related to the dimensional constants μ_0 and ε_0. Thus, the inverse-square laws of electricity and magnetism are, respectively,

(3.11)
$$F = \frac{Q_1 Q_2}{4\pi \varepsilon_0 r^2}$$

and

(3.12)
$$F = \frac{Q_{m1} Q_{m2}}{4\pi \mu_0 r^2}.$$

This symmetry, or duality, is not created by the insertion of μ_0 into eqn (3.9), but it is highlighted by that equation. The underlying physical reasons will become clearer in Chapter 5 when the mutal relationships between electric and magnetic energy are discussed.

3.3 Tubes and slices, inductance, and boundary conditions

Ampère's equivalence theorem enables the ideas of electrostatics to be applied to the interaction of steady currents in terms of magnetic fields. We define tubes of magnetic flux, Φ, and a magnetic flux density, B, where

(3.13)
$$\Phi = \int B_n \, dS \quad (\text{Wb}).$$

Since there are no free magnetic poles, note that

(3.14)
$$\oint B_n \, dS = 0.$$

We can also define slices of magnetic potential, V^*, and the magnetic field strength, H, as a gradient of the potential

(3.15)
$$\int_1^2 -H_l \, \mathrm{d}l = V_2^* - V_1^* \quad \text{(A)}.$$

In terms of the pole strength, H is the force per unit pole and

(3.16)
$$H = \frac{Q_m}{4\pi\mu_0 r^2} \quad \text{(A m}^{-1}\text{)}$$

The fact that the magnetic potential, V^*, has units in amperes can be deduced from the fact that the unit of flux is the weber or volt second and the unit of energy is the joule, which is given by the product of potential and flux in a tube or slice. The energy density is

(3.17)
$$\frac{1}{2}\mu_0 H^2 = \frac{1}{2}\frac{B^2}{\mu_0} = \frac{1}{2} HB \quad \text{(J m}^{-3}\text{)},$$

and the tensile stress in a tube is

(3.18)
$$\tfrac{1}{2}HB \quad \text{(N m}^{-2}\text{)}.$$

Polarizable magnetic materials can be described in terms of the relative permeability, μ_r, where

(3.19)
$$\mu_r = 1 + \chi_m,$$

where χ_m is the magnetic susceptibility.

The parameter analogous to the capacitance is the inductance, it is given by

(3.20)
$$L = \frac{Q_m}{V^*} = \frac{\Phi}{V^*} \quad \text{(Wb A}^{-1} \text{ or H)}.$$

We shall redefine the inductance in terms of current loops in Section 5.6 when we discuss the implications of Ampère's equivalence theorem more fully. The definition in eqn (3.20) can be applied to problems outside a region of current.

It is interesting to note that the units of the product $(LC)^{1/2}$ are seconds and $(L/C)^{1/2}$ has units in ohms.

The boundary conditions at an interface between a polarizable magnetic material of relative permeability μ_r are analogous to the boundary conditions at the surface of a dielectric which were discussed in Section 2.11. The normal component of the magnetic flux density, B, is continuous, and so is the tangential component of the magnetic field strength, H, if there is no surface current. These conditions are illustrated in Fig. 3.3.

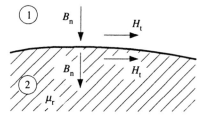

Fig. 3.3 Magnetic boundary conditions.

It is often useful to express the surface polarity, σ_m, in terms of the applied normal field at the surface. The polarity will itself produce a field, as illustrated in Fig. 3.4. By using the continuity of B_n and noting that B_n in Fig. 3.3 will induce a negative σ_m, we have for an applied field H_0 normal to the surface and directed outwards

(3.21)
$$\mu_0\left(H_0 + \frac{\sigma_m}{2\mu_0}\right) = \mu_r\mu_0\left(H_0 - \frac{\sigma_m}{2\mu_0}\right),$$

(3.22)
$$\sigma_m = 2\mu_0 \frac{\mu_r - 1}{\mu_r + 1} H_0.$$

Notice that, for large values of the relative permeability, σ_m is almost independent of μ_r, and is proportional to the applied normal field at the surface. Notice also that a smooth surface is assumed, which gives the effect of being flat locally. Equation (3.22) does not apply for the polarity near a corner of the material. This combination of the field components gives the boundary conditions illustrated in Fig. 3.5, where $\tan \theta_2 = \mu_r \tan \theta_1$. Since μ_r in iron is of the order of 1000, this means that $\theta_1 \ll \theta_2$ and it can be deduced that the magnetic field just outside an iron surface is approximately perpendicular to that surface.

Inside the material, the field is given by H in a long narrow cylinder and by $\mu_r H$ in a short wide cylinder, if the cylinders have their axes along the

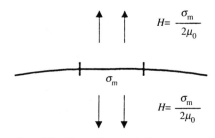

Fig. 3.4 The field produced by the surface polarity.

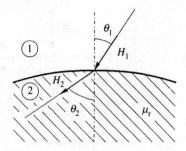

Fig. 3.5 Boundary conditions.

polarization. See Section 2.10 and particularly Fig. 2.28 which discusses the analogous electrical case.

3.4 Magnetic shells

Ampère's equivalence theorem makes use of small current loops and dipoles. In this context, small means small compared with the distance at which the field is observed using a probe which is either a dipole or a current loop. There is no difficulty in applying this idea to dipoles, because a large magnet can always be subdivided into arbitrarily small dipoles. The distance between the two poles of such a dipole can be reduced and the dipole moment can be kept constant by increasing the pole strengh correspondingly. But there is a difficulty in dealing with current loops because they necessarily have a hole which defines their shape as a loop. The conductor forming the loop has to have a diameter which is small compared with the loop diameter and it is impossible to maintain the shape if both the area of the loop and the diameter are to be negligibly small. It is therefore difficult to apply the equivalence between current loops and dipoles in the vicinity of current-carrying conductors.

Ampère overcame this difficulty in a very ingenious way. First, he subdivided a large current loop into small loops as shown in Fig. 3.6.

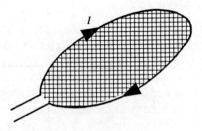

Fig. 3.6 A subdivision of a large current loop.

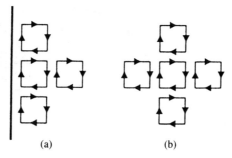

Fig. 3.7 (a) Edge and (b) internal loops.

Figure 3.7 shows an enlarged view of the small loops. Two possibilities arise: either the small loop is adjacent to the external perimeter, as in Fig. 3.7(a), or it is not, as in Fig. 3.7(b). In the case of Fig. 3.7(b), all the currents of the central small loop are cancelled by those of neighbouring loops; but, in the case of Fig. 3.7(a), the current of the piece of the original large loop is not cancelled. Hence the addition of all the loops is equivalent to the original loop. A loop of arbitrary size can therefore be represented by a distribution of small loops over a surface which is bounded by the large loop. Furthermore, this distribution of small loops is not uniquely confined to a particular surface. All that is required is that it should be bounded correctly at the perimeter.

Suppose now that we wish to examine the magnetic field at any arbitrary point near a large current loop. The large loop can be replaced by a surface distribution of small loops, taking care to choose a surface at a considerable distance from the point of interest. Each small loop of that surface can now be represented by a dipole; and the entire surface can be transformed into a double layer of magnetic pole strength, one side of the surface being positive and the other negative. This surface layer of diples is called a magnetic shell. This idea makes it possible to represent current loops of arbitrary size by equivalent distributions of magnetic polarity.

3.5 Magnetomotive force

Figure 3.8 shows a current loop and one of its equivalent magnetic shells. From the point of view of the magnetic field these two objects are identical, but from the point of view of energy conversion they are very different. The difference is illustrated in Fig. 3.8 by the fact that the current loop has external connections and the magnetic shell is an isolated object. Energy can be exchanged with the current loop by altering the current supply, but this is impossible in the case of the shell. Nevertheless, the magnetic field at any arbitrary external point is identical. This apparent paradox needs to be

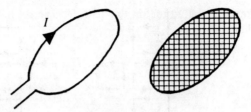

Fig. 3.8 A current loop and a magnetic shell.

examined. The key to it lies in the idea of a conservative field, which was met in the context of electrostatics.

In such a field, every point can be associated with a potential energy and therefore no change of this energy is associated with any closed path. This is true for any field which can be represented by the flow of an incompressible fluid and therefore applies to the field of electric charges and magnetic poles. But, for the magnetic field of currents, it is true only if the current has been replaced by a magnetic shell. That replacement, as we have seen, requires the use of a shell at a distance from the point of interest. Suppose now we want to examine all points around a current loop at the same instant. Then it becomes impossible to choose a magnetic shell which is distant from all these points if points are chosen on a path which goes through a magnetic shell, or links the current loop.

Consider a path through a magnetic shell, which can be considered flat locally, as illustrated in Fig. 3.9. The pole strength per unit area is q_m. For any closed path the field of a set of poles is conservative, so that

(3.23)
$$\oint H_l \, dl = 0.$$

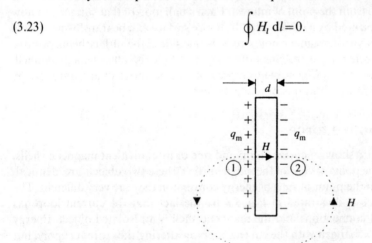

Fig. 3.9 A magnetic shell.

From Fig. 3.9, note that the field, H, reverses inside the shell. The internal field is given by

(3.24)
$$H = \frac{q_m}{\mu_0}$$

and

(3.25)
$$\int_1^2 -H_l \, dl = -\frac{q_m d}{\mu_0}.$$

Hence the external contribution is, by the use of eqn (3.23),

(3.26)
$$\int_2^1 -H_l \, dl = +\frac{q_m d}{\mu_0}.$$

The work done externally is replaced internally.

Next, consider a path linking the current loop. Here the external contribution is the same, but there is no internal replacement, because there is no local magnetic shell. Moreover, the distance d can be made infinitesimally small so that for the current loop

(3.27)
$$\oint -H_l \, dl = \frac{q_m d}{\mu_0}.$$

Noting that $q_m = Q_m/S$, then using eqn (3.9)

(3.28)
$$\oint H_l \, dl = I.$$

For a path linked with the current the field is no longer conservative as it was in eqn (3.23). Work is done by a unit pole traversing a closed loop, and there is an exchange of energy with the source which supplies the current, I. The expression on the left-hand side of eqn (3.28) is called the *magneto-motive force* (m.m.f.), and its unit is the ampere.

3.6 The magnetic circuit and permeance

It was shown in Section 3.3 (see eqn 3.14) that the net magnetic flux out of any closed surface is zero. In this respect, magnetic flux resembles a steady electric current in a conductor. Indeed, there can be no net outflow of current from any closed surface. This is consistent with the assumption of steady conditions which does not permit any change of electric charge within the surface.

This similarity gives rise to the idea known as the *magnetic circuit*. It is based on the analogy between the circulation of magnetic flux in a closed path and the circulation of electric current in a closed circuit. This analogy yields some very simple and quite accurate calculations of inherently complicated magnetic-field problems. The procedure is illustrated in Fig. 3.10. A coil may be used to magnetize an iron core. The iron forms a closed path and the flux will circulate around this magnetic circuit as indicated. Assuming there is no magnetic saturation, the leakage of flux into the surrounding air will be small, and thus the total flux is practically the same over all cross sections of the core. This resembles an electric circuit where the current is the same over all cross sections of the wire. The m.m.f. of the coil (as given by its ampere-turns, In, where I is the supply current and n is the number of turns of the coil) could be considered to be the analogue of the e.m.f. of the battery in an electric circuit. The flux in a magnetic circuit is then the analogue of the current in an electric circuit. Equation (3.28) can be applied in much simpler form with the integration replaced by a summation of terms with H constant in each segment of the closed magnetic path

(3.29)
$$\sum_i H_{li} l_i = In,$$

which resembles going round the loop in an electric circuit.

The magnetic analogue of electrical resistance is called the *reluctance*. For any part of the magnetic circuit, the reluctance is defined in terms of the magnetic potential and flux by

(3.30)
$$R_\mu = \frac{V_1^* - V_2^*}{\Phi} = \frac{1}{\Phi} \int_1^2 H_l \, dl$$

and for the complete circuit

(3.31)
$$\text{reluctance} = \frac{\text{m.m.f.}}{\text{flux}} = \frac{In}{\Phi}.$$

The units of reluctance are ampere per weber (A Wb^{-1}). The reciprocal of reluctance is called the *permeance*.

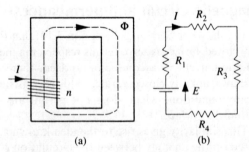

(a)   (b)

Fig. 3.10 (a) A magnetic circuit, and (b) an electric-circuit analogue.

By analogy with Ohm's law in electric circuits

(3.32)
$$\text{e.m.f.} = IR$$

and thus

(3.33)
$$\text{m.m.f.} = \Phi R_\mu$$

where, for simple parallel-sided segments of iron core, like those in Fig. 3.10,

(3.34)
$$R_\mu = \frac{l}{\mu S},$$

where l is the length, S is the cross-sectional area, and μ is the permeability. This last equation closely resembles the expression, eqn (1.2), for the resistance of parallel-sided conductors (see Section 1.2) if μ is substituted for σ (conductivity).

The magnetic-circuit analogy is often used in approximate calculations of the magnetizing current which is required to set up a given magnetic flux in the iron circuit of an electrical device. We shall explain the procedure with reference to the very simple case illustrated in Fig. 3.11. The presence of a small air gap is quite common, and in some devices it is essential so that there may be relative motion between the different parts. From eqn (3.29)

(3.35)
$$\sum H \, \mathrm{d}l = H_{\text{iron}} l_{\text{iron}} + H_{\text{gap}} l_{\text{gap}} = In$$

so that

(3.36)
$$nI = \frac{B l_{\text{iron}}}{\mu_r \mu_0} + \frac{B l_{\text{gap}}}{\mu_0} = \frac{B l_{\text{gap}}}{\mu_0} \left(1 + \frac{l_{\text{iron}}}{\mu_r l_{\text{gap}}} \right).$$

Given the actual dimensions, the permeability of iron, and the required flux density in the air gap then the ampere-turns, and thus the magnetizing current, may be easily calculated. Suppose $\mu_r = 3000$ and $l_{\text{iron}} = 30 l_{\text{gap}}$. Then

(3.37)
$$nI = \frac{B l_{\text{gap}}}{\mu_0} (1 + 0.01),$$

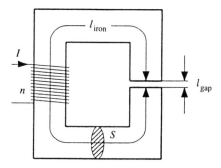

Fig. 3.11 A magnetic circuit with an air gap.

which shows that only 1 per cent of the m.m.f. is absorbed by the iron path, although it is so much longer than the air gap; 99 per cent of the m.m.f. is used to force the flux across the air gap.

The principle of the calculation is very easy, but in practice it often becomes modified and refined. First, the analogy between magnetic and electric circuits is not complete because magnetic flux and electric current are very different entities. Electric current is formed by a flow of electrons, whereas nothing flows when there is a magnetic flux. Thus, energy is required to set up a magnetic field, but none is required to maintain it; otherwise permanent magnets could not exist. Another imperfection of this analogy arises from the fact that the ratio of the conductivities of a typical conductor and an insulator is immensely greater than the ratio of the permeabilities of iron and air. Thus, air or free space becomes a legitimate, though not a very good, 'conductor' of flux (usually in the form of air gaps like the one in our example); at the same time the possibility of leakage flux must be kept very much in mind. This problem is considered in the next section. And last, but not least, Ohm's law for electric circuits tells us that resistance is independent of current, whereas the reluctance in magnetic circuits is not independent of the flux because the permeability varies with Φ. The *magnetic saturation* may come into consideration, changing the proportions suggested by Eqn (3.37).

3.7 Calculation of permeance using tubes and slices

In many electrical devices, magnetic circuits are designed in such a way that very little m.m.f. is absorbed in the iron core and attention is focused on the shape and dimensions of the air gap (refer to the example in the previous section). The core provides a path for transporting the magnetic flux from the place it is produced (the current in the coil) to the place where it can be used (the air gap), which may be well away from the winding. An unsaturated iron surface may be assumed to have a constant magnetic potential and thus it becomes a slice. The whole air-gap region may be enclosed using a pair of slices (iron surfaces), and at the same time the flux distribution may be described in terms of tubes. The tubes terminate on iron surfaces. Yet again the now familiar pattern of tubes and slices can be recognized with its computational advantages. For convenience, we shall work in terms of the permeance (which has already been introduced as the reciprocal of the reluctance). Permeance in magnetic circuits is the analogue of conductance in electric circuits. A parallel-sided tube of flux, of length δl and cross section δS, may therefore be seen to have a permeance given by

(3.38)
$$\Lambda = \mu \frac{\delta S}{\delta l}.$$

If there are m such pieces along a tube, then these permeances are in series and are given by

$$\frac{1}{\Lambda} = \sum_1^m \frac{\delta l}{\mu \, \delta S}.$$

(3.39)

If there are n tubes in parallel, the total permeance is

$$\Lambda = \sum_1^n \left(\sum_1^m \frac{\delta l}{\mu \, \delta S} \right)^{-1}.$$

(3.40)

If the slices are considered instead, then for n slices in parallel

$$\Lambda = \sum_1^n \frac{\mu \, \delta S}{\delta l}$$

(3.41)

and for m slices in series

$$\Lambda = \left[\sum_1^m \left(\sum_1^n \frac{\mu \, \delta S}{\delta l} \right)^{-1} \right]^{-1}.$$

(3.42)

Equations (3.39) to eqn (3.42) are similar to eqns (2.20) to eqn (2.23) for calculating the capacitance and eqns (3.38) to eqn (3.42) are similar to eqns (1.11) to eqn (1.15) for calculating the resistance. The analogue of the permeability, μ, is the permittivity, ε, or the conductivity, σ. The subdivision into tubes produces the lower bound, whereas slices give the upper bound of the permeance. If there are no currents in the region of interest the analogy with electric fields is complete, except for the proviso that there are no free magnetic poles. This means that flux tubes are closed. Nevertheless, interest often centres on the magnetic field in a particular region such as the air gap of a machine, so that only a part of the flux tubes needs to be examined. These tubes will typically be assumed to terminate on iron surfaces, as suggested before.

Furthermore, the definition of *inductance* was introduced in terms of the flux and the magnetic potential for problems outside regions containing a current (eqn 3.20). Comparison with eqn (3.29) shows that in such problems the calculation of inductance is reduced to a calculation of permeance. In eqns (3.38) to eqn (3.42) L may be substituted for Λ. We now have a system of equations for calculating the circuit parameters, R, L, or C, for many practical problems under static conditions. For fuller information, specialist books and papers should be consulted (see the Bibliography in Appendix 5).

The problem of leakage flux, already mentioned in the previous section, deserves some consideration. This effect will be illustrated with reference to Fig. 3.11. In our simplified calculations the flux was assumed to be the same over all cross sections of the iron, and the flux density in the air gap was assumed to be uniform and the same as in the core. A more realistic picture is that some of the flux will pass directly through the air inside or outside the

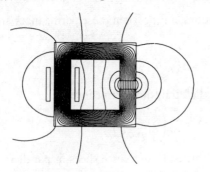

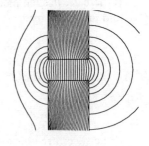

Fig. 3.12 Leakage flux.   **Fig. 3.13** Fringing around the air gap.

core. These effects are known as the *leakage* and *fringing* flux, and are illustrated in Figs 3.12 and 3.13, respectively. Such leakage and fringing fluxes may be conveniently studied using the tubes-and-slices approach.

3.8 The magnetic field of a long cylindrical conductor

Ampère's equivalence allows all steady currents to be represented by magnetic shells; this is often a great advantage because is allows the methods of electrostatics to be applied to magnetostatics. However, this equivalence only applies to regions which are bounded by magnetic shells, and therefore it forbids the use of closed regions linking the current. Problems of energy interchange between the magnetic field of currents, and the sources of those currents, require the use of the magnetomotive force instead of magnetic potentials.

The magnetomotive force in eqn (3.28) does not define the path by which the current is linked, nor does it define the local values of H, which depend on that path. In general, therefore, it does not make it possible to predict the local value of H. Magnetomotive force is a *global* rather than a *local* property. In our discussion of Gauss's theorem we met a similar problem. Gauss's theorem deals with the total flux over a closed surface and not with the local flux density. Global properties are described in terms of closed integrals which depend on the shape of a physical object. In Gauss's theorem the surface encloses the sources, and in the circuital law of magnetostatics the line links with the current.

The importance of shape draws attention to the idea of symmetry. If every part of a surface or every part of a line is indistinguishable from every other part then the local fields will also be indistinguishable. For Gauss's theorem,

(3.43)
$$\oint D_n \, dS = D_n S,$$

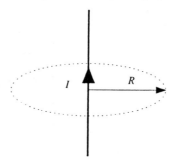

Fig. 3.14 A current filament.

and for the circuital law,

(3.44)
$$\oint H_l \, dl = Hl.$$

Consider, for example, the magnetic field of a long straight current of small cross-sectional area (Fig. 3.14). This is known as a *current filament*. A circular path around the current is symmetrical, thus

(3.45)
$$\oint H_l \, dl = H(2\pi R) = I.$$

Hence

(3.46)
$$H = \frac{I}{2\pi R}.$$

Notice that this is the only component of H due to the current. Strictly speaking this result applies only to an isolated current of infinite length. Physically, the current must have a return somewhere and it must also have ends. But eqn (3.46) is reliable as long as the length of the current is large compared to R and the return conductor is at a distance that is large compared to R.

If the current were to flow in a filament of zero diameter, the magnetic field close to it would become infinite and the local energy density would also be infinite. Moreover, the current density would be infinite. In a resistive conductor this would mean that the potential difference driving the current would have to be infinite. In practical terms, the conductor would melt and disintegrate. Equation (3.46) applies to the outside of currents of finite cross-sectional area.

Now consider a long current uniformly distributed over the cross section of a cylinder of radius a (Fig. 3.15). At an internal radius R,

(3.47)
$$\oint H_l \, dl = H(2\pi R) = I \frac{R^2}{a^2},$$

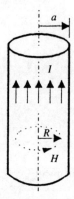

Fig. 3.15 A cylindrical conductor.

(3.48)
$$H = \frac{IR}{2\pi a^2}.$$

The graph of the internal and external magnetic field of the current is given in Fig. 3.16.

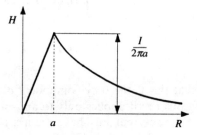

Fig. 3.16 The magnetic field strength of cylindrical conductor.

3.9 The magnetic field of a current element

The magnetic field in the neigbourhood of a current depends on the entire current, but it would be helpful in calculations if the entire current system could be treated as being made up of a succession of short current elements, each of which contributes to the magnetic field. The total field could then be found by numerical addition or integration. This idea is illustrated in Fig. 3.17. The magnetic field at a point P due to a current element $I\,dl$ at a distance r is required.

There is, however, a difficulty in this apparently simple procedure. If a formula is devised, it could not be tested by measurement because, physically, there is no such thing as a current element carrying a steady

Fig. 3.17 A current element.

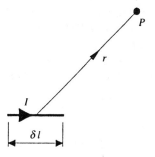

current. The current must flow in a loop or a circuit and a current element cannot be isolated.

This difficulty was overcome in an ingenious way by Oliver Heaviside, who proposed a 'rational' current element. To enable the current to flow, he suggested that the current element should consist of a short piece of an insulated conductor immersed in a large volume of a conducting liquid. A suitable potential difference could be applied between the ends of the conductor and the current could flow in at one end and out at the other as illustrated in Fig. 3.18. If the element has a negligible cross section, the inflow and outflow would be symmetrically disposed like the field around a small spherical conductor. The circuital law could then be used to calculate the field at P as illustrated in Fig. 3.19. Consider first the outflow over the

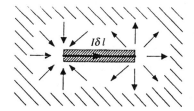

Fig. 3.18 Heaviside's 'rational' current element.

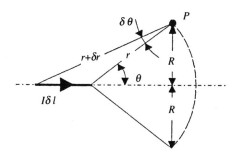

Fig. 3.19 The magnetic field of a current element.

spherical cap of radius r. The current density is $I/4\pi r^2$, and the area of such a cap is $2\pi r^2(1-\cos\theta)$. Hence the outflow is $I(1-\cos\theta)/2$. The inflow over the same cap is $I(1-\cos|\theta-\delta\theta|)/2$. Hence the net outflow is $I(\cos|\theta-\delta\theta|-\cos\theta)/2 = (I\delta\theta\sin\theta)/2$. Now applying the circuital law and using the symmetry of Fig. 3.19, then

(3.49)
$$\oint H_t\,dl = H(2\pi R) = \tfrac{1}{2}I\,\delta\theta\sin\theta.$$

Since $R = r\sin\theta$ and $r\,\delta\theta = \delta l\sin\theta$,

(3.50)
$$H = \frac{I\,\delta l}{4\pi r^2}\sin\theta.$$

Notice that the field is perpendicular to the plane containing δl and r, and it is related to the direction of the current by the motion of a right-handed screw.

A real circuit would be made up of a succession of elements, as illustrated in Fig. 3.20. At the junction between successive elements, the outflow and inflow would cancel each other, and, finally, when the current circuit is complete all the current would flow in the conductor. Then the conducting liquid could be removed. Hence, application of eqn (3.50) would give the correct answer for the magnetic field without any reference to the liquid.

Heaviside's current element is not only a clever device to be used in calculation, but it also has a deeper physical significance. It will have been noticed that only steady currents have been dealt with so far. Ampère's experiments were all on steady currents and the circuital law applies to such currents. But the treatment is extended to alternating currents later in this book. Then the energy distribution in space acts rather like Heaviside's conducting liquid and an expression can be derived for the magnetic field without having to use a hypothetical liquid. Equation (3.50) is then found to be the low-frequency approximation to the magnetic field of an alternating-current element.

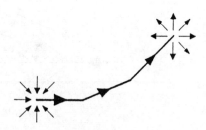

Fig. 3.20 A succession of current elements.

3.10 The force on a current in a magnetic field

Equation (3.50) enables us to calculate the force acting on magnetic dipoles in the vicinity of a current element and its conducting liquid. Since the force on a dipole is the resultant of the forces on its ends, the discussion can be simplified by considering the force on an isolated pole, as illustrated in Fig. 3.21. The force is

(3.51)
$$HQ_m = \frac{I\,\delta l \sin\theta\, Q_m}{4\pi r^2}.$$

By Newton's law of action and reaction there will be an equal and opposite force on the current element. Moreover this force will act entirely on the element and it will not act on the current in the liquid, because that current has spherical symmetry. Also, it is possible to measure the force on an element of a current circuit by giving it suitable flexible connections. Now the flux density, B, of the pole, Q_m, is given by

(3.52)
$$B = \frac{Q_m}{4\pi r^2}.$$

Hence the reaction force on the current element is given by

$$I\,\delta l \sin\theta\, B$$

into the paper. But $B\sin\theta$ is the component of B perpendicular to δl. So the force is

(3.53)
$$F = I\,\delta l B_{\perp}$$

and its direction is obtained by turning a right-handed screw from the δl direction into the $B_{\perp}$ direction. We have now arrived back at eqn (3.1), which defined the unit of current. From eqn (3.46), the field of a long current filament is given at a distance a by

(3.54)
$$B = \frac{\mu_0 I}{2\pi a}.$$

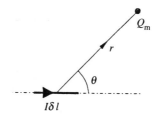

Fig. 3.21 A calculation of the force on an isolated pole.

Hence the force on a parallel current per unit length is given from eqn (3.53) by

(3.55)
$$F = \frac{\mu_0 I_1 I_2}{2\pi a}.$$

We have used Ampère's experimental results, which led to the equivalence of current loops and dipoles. Secondly, we have used the fact that the fields of dipoles are conservative. Thirdly, we have used Newton's law of action and reaction. Our argument has shown that the kinetic energy of steady currents can be expressed in terms of a magnetic field with an energy density of $\frac{1}{2}HB$ distributed in space in the vicinity of an electric current or dipoles.

3.11 Stress in a magnetic field

It has already been seen that tubes of magnetic flux carry a tensile stress of $\frac{1}{2}HB$ (see eqn 3.18). This is similar to the tensile stress of $\frac{1}{2}ED$ carried by the electric flux. At first sight, it appears strange to associate stress and force with a flux acting in space which may be devoid of matter. But, on reflection, it will be seen that these stresses are strictly analogous to mechanical stresses. Suppose a steel rod is under tension, as illustrated in Fig. 3.22. The stress at a particular section is obtained by considering the rod to be cut at that section, as in Fig. 3.23, and considering one part of it to have been removed. Then for equilibrium there must be a force, P, applied at that section, and the stress can be calculated by dividing the force by the area. The stress in a tube of flux is obtained in the same manner. Consider the tube shown in Fig. 3.24. The tube must be 'cut' or terminated at the

Fig. 3.22 A steel rod under tension.

Fig. 3.23 A rod cut to find the stresses.

Fig. 3.24 A tube of magnetic flux.

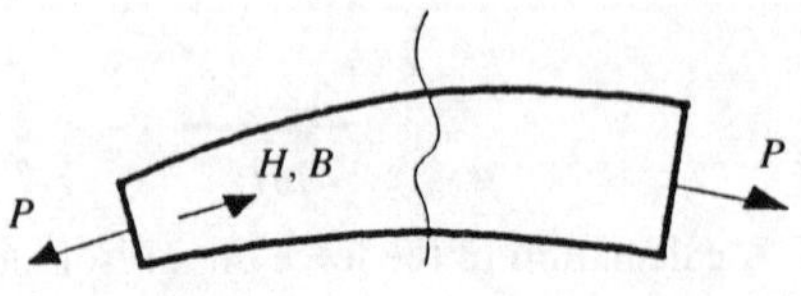

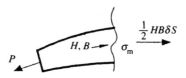

Fig. 3.25 A cut in a tube of flux.

section under consideration. This means that a suitable layer of magnetic pole strength must be placed there, as in Fig. 3.25. The pole strength is given by

(3.56)
$$\sigma_m = -B.$$

The stress on σ_m from the field to the left is $\frac{1}{2}Hq_m = -\frac{1}{2}HB$, and for equilbrium to be maintained an equal and opposite stress must be applied at the free surface. The factor $\frac{1}{2}$ again arises from the fact that the average field at the surface is $\frac{1}{2}H$. To put this another way, the field of σ_m by itself is as shown in Fig. 3.26. So the external force is given by $\frac{1}{2}H$. It can be seen that the idea of a field stress always implies the insertion of a suitable pole strength to terminate the tube of flux. It does not imply a force in empty space.

The tensile stress is not sufficient to maintain equilibrium between tubes. Consider Fig. 3.27. To consider an isolated tube, the adjacent tubes need to be removed. This can be done by surrounding the tube with a suitable current. It has been seen that the pole strength terminates the normal component of the magnetic field. Similarly, a surface current can terminate the tangential component, as shown in Fig. 3.28, where the current is out of the paper at the top of the tube and into the paper at the bottom.

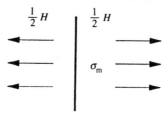

Fig. 3.26 A magnetic field of pole strength q_m.

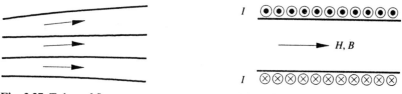

Fig. 3.27 Tubes of flux.

Fig. 3.28 A surface current.

To simplify the problem, consider first a two-dimensional field which has no variation in the direction perpendicular to the paper, as in Fig. 3.29. Consider a loop of length l close to the boundary, and apply the circuital law. If there is no field above the current and there is a field H below, then

(3.57)
$$Hd = Jd,$$

where J is the line density of the current (in A m^{-1}). Hence a line density of J (in A m^{-1}) terminates a tangential field H (A m^{-1}). We know that the force on a current in a magnetic field is given by

(3.58)
$$F = IlB_\perp.$$

This force is perpendicular to l and $B_\perp$. In this example, l is upward and B is from left to right. This gives a force on the current which is in the direction indicated. This force is

(3.59)
$$F = \tfrac{1}{2}IlB,$$

and the stress is force/area and it is given by

(3.60)
$$\tfrac{1}{2}JB = \tfrac{1}{2}HB.$$

This expression is not limited to the two-dimensional case because stress is associated with a small area, which can always be assumed to be flat.

For equilibrium there must be an equal and opposite stress, that is, a compressive stress on the tube. We now have a complete stress system: a tensile stress, $\tfrac{1}{2}HB$, along the tube and a compressive stress, $\tfrac{1}{2}HB$, perpendicular to the tube surface. This stress system provides a useful tool for calculating both magnetic fields and electric fields. For the latter, a compressive stress of $\tfrac{1}{2}ED$ can be inferred on a tube surface by using the duality between electric and magnetic fields. We did not discuss the compressive stress in the chapter on electrostatics because we did not wish to invoke a 'magnetic current' as a boundary to an electric tube, but since this current does not appear in the expression for stress it does not matter. It would be difficult to isolate an electric tube experimentally without a

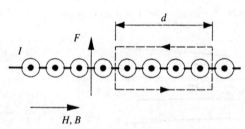

Fig. 3.29 The termination of a tangential magnetic field.

magnetic current, although we shall find that a changing magnetic flux has the same effect. It would also be difficult to terminate a magnetic tube at its ends, because there are no free magnetic poles. But an iron surface gives a good approximation because the internal H can be made very small. In any case, field stresses are a calculating tool. Discussion of the behaviour of isolated tubes is in the realm of thought experiments rather than laboratory experiments.

Exercises

3.1 Discuss the equivalence of a small current loop and a small magnetic dipole. Why must the effect be observed from a large distance and what is meant by the word large in this context? State the inverse-square law for magnetostatics and compare it with a similar expression for electrostatics.

3.2 Define *tubes* of magnetic flux and *slices* of magnetic potential, and relate them to the field quantities B (the flux density) and H (the magnetic field strength). Discuss the analogy with similar definitions in electrostatics. Define the terms: *permeability*, *susceptibility*, and *inductance*, and explain which electrostatic quantities are analogous.

3.3 By analogy with electrostatics, define magnetic boundary conditions between regions of different permeability. Why is the magnetic field just outside an iron surface approximately perpendicular to that surface?

3.4 What is a *magnetic shell*? Why is the concept of a magnetic shell useful? How does the equivalence of a current loop and a magnetic shell lead to the idea of a *magnetomotive force* (m.m.f.)?

3.5 Define the terms *reluctance* and *permeance*. Explain how the concept of a *magnetic-circuit analogy* is helpful in approximate calculations in magnetostatics. Specify the magnetic circuit equivalents of the following electric circuit quantities: *e.m.f.*, *current*, *current density*, and *resistance*. Explain why the analogy between magnetic and electric circuits is not complete.

3.6 A mild-steel ring having a cross-sectional area of 10 cm^2 and a mean circumference of 60 cm has a coil of 300 turns wound uniformly around it (see Fig. 3.30). The B/H curve for mild steel may be approximated using the formula

$$B = \frac{H}{192.5 + 0.57H}.$$

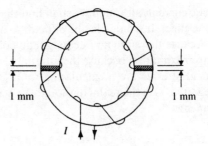

Fig. 3.30 The magnetic field in a ring.

Calculate the current required to produce a flux of 1.6 mWb in the ring.
(*Answer* 7 A.)

3.7 If the mild-steel ring of Exercise 3.6 is cut radially at diametrically opposite points and a sheet of brass 1 mm thick inserted in each gap, as indicated in Fig. 3.30, find the current required to produce a flux density of 1.6 T. Assume that there is no magnetic leakage of fringing, that is, that all flux passes round the iron core and directly across the brass strips. The relative permeability of brass is $\mu_r \approx 1$.
(*Answer* 15.5 A.)

3.8 A model of a simple actuator mechanism is shown in Fig. 3.31. The iron yoke is fixed and the plunger is free to move horizontally. All the components have the same cross-sectional area, S. Both the yoke and the plunger are infinitely permeable. Assuming that leakage and fringing are negligible, derive an expression for the inductance of the coil.
(*Answer* $L = \mu_0 N^2 S/2x$.)

3.9 In magnetically linear systems the force acting in a particular direction may be found as

$$F_x = \tfrac{1}{2} I^2 \frac{\partial L}{\partial x}.$$

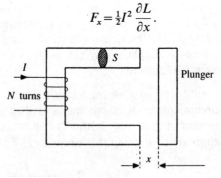

Fig. 3.31 A simple actuator.

Derive a formula for the force acting on the plunger in the actuator mechanism of Fig. 3.31.

(*Answer* $F = \mu_0(NI)^2 S/4x^2$ and it is attractive.)

What happens as $x \to 0$?

3.10 Discuss the use of *tubes and slices* in calculations of dual bounds of a permeance (inductance). Discuss the analogy of such calculations with the estimation of the dual bounds of resistance or capacitance.

3.11 In order to study the effect of fringing around an air gap (see Fig. 3.13), the two-dimensional model in Fig. 3.32 is assumed. Use the tubes-and-slices method to estimate the change in the inductance due to an increased air gap.

Solution

If L_0 is taken as the value of the inductance with fringing neglected, where

$$L_0 = \frac{\mu_0 N^2 w}{d}$$

per unit depth, then the actual inductance with fringing included, L, may be plotted as in Fig. 3.33.

Fig. 3.32 Fringing around the air gap.

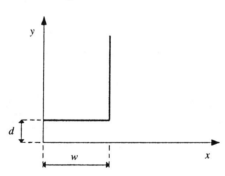

Fig. 3.33 The change in L due to the air gap.

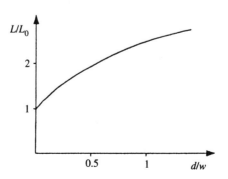

3.12 Estimate upper and lower bounds of the inductance per unit length of the coaxial line with a strip centre-conductor, as illustrated in Fig. 3.34.

Solution
Due to symmetry, an appropriate computational model may be assumed as shown in Fig. 3.35(a). Distributions of tubes and slices which are suitable for hand calculations are shown in Fig. 3.35(b) and Fig. 3.35(c), respectively. A typical distribution obtained from the tubes and slices (TAS) program (see Appendix 1) is demonstrated in Fig. 3.36. A finite-element solution is shown in Fig. 3.37. The results are summarized in Table 3.1.

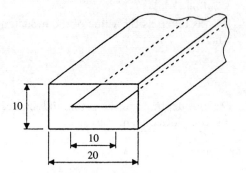

Fig. 3.34 A rectangular coaxial line.

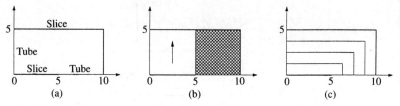

Fig. 3.35 (a) The model, (b) tubes, and (c) slices.

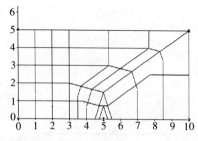

Fig. 3.36 TAS solution.

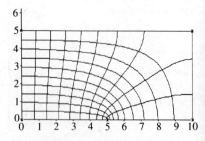

Fig. 3.37 Finite-element solution.

Table 3.1. Summary of results

Method of calculation	Inductance per unit depth ($\times \mu_0$)
Tubes-and-slices hand calculation	1.42
Tubes-and-slices (TAS program)	1.45
Finite elements (TAS program)	1.47

3.13 In an experiment, it is required to 'screen' a region from a constant magnetic field. Two iron plates are available for this purpose. Should they be placed across the magnetic field or along it? Sketch the resulting flux distribution in either case.
(*Answer* along the field.)

3.14 Figure 3.38 shows a conductor carrying a current I in a slot of width d cut in iron of very high relative permeability. Using the method of magnetic-field stresses show that the downwards force on the current is given by

$$F = \frac{1}{2}\frac{\mu_0 I^2}{d}.$$

3.15 Figure 3.39 shows a simple torque motor. If the exciting ampere-turns are NI, both air gaps have length g, the iron reluctance is negligible, the rotor diameter is d, and the axial length of the machine is l, estimate the torque using the method of field stresses ($d \gg g$).
(*Answer* $\frac{1}{8}\mu_0 \, N^2 I^2 l d/g$ (N m).)

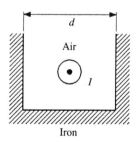

Fig. 3.38 The force on the current in a slot in iron.

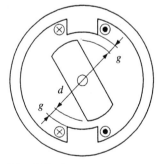

Fig. 3.39 A torque motor.

Electric and magnetic fields as vectors | 4

4.1 Introduction to vectors

In the previous three chapters electric and magnetic phenomena were described in terms of tubes of flux associated with slices of potential differences. The tubes and slices cut each other at right angles, and at every intersection a suitable system of right-angled coordinates could be constructed (Fig. 4.1). But such coordinates would vary from place to place, so that it would not in general be possible to determine the direction of the field from a single set of coordinates. Instead of using a single quantity like E the three components of E have to be specified, say E_x, E_y, and E_z, as projections of the vector E on to the axes of an x, y, z coordinate system. The vector would need to be specified by three numbers. Computationally there is no great difficulty in handling three numbers rather than one, but the use of the projections of a vector rather than the vector itself makes it difficult to understand the physical relationships in the field. The projections have no physical significance because the choice of coordinates is in the hands of the analyst and, except in very symmetrical systems, this choice is arbitrary. Indeed, one of the chief means of clarifying electromagnetic behaviour has been the desire to rid the description of arbitrary coordinate systems. This is the motive behind the development of an algebra of vectors in this chapter. We shall obtain not only mathematical simplicity from the introduction of vectors but also physical insight.

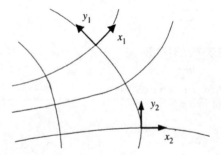

Fig. 4.1 Intersection of tubes and slices.

4.2 The gradient vector

In the description of tubes and slices, we have to deal with two kinds of vectors. Consider the slices first. Here there are equipotential surfaces and a vector, which is the potential gradient, as illustrated in Fig. 4.2. The gradient is perpendicular to the equipotential surface, because by definition there is no gradient along such a surface. In order to define a local gradient two equipotentials close to each other need to be considered, so that they can be considered to be locally parallel to each other, as in Fig. 4.3. Then the gradient can be written as

(4.1)
$$\text{gradient } V = \frac{\partial V}{\partial n} \, \hat{n},$$

where $\hat{n}$ is a unit vector in the direction perpendicular to the equipotentials, and $\partial V/\partial n$ is the magnitude of the gradient. The gradient of V is a vector having both magnitude and direction. The operation of taking the gradient of a potential converts that potential, which is a single number (a scalar), into a vector. It is useful to regard the word gradient as an operator; a gradient must be the gradient of a potential. If the potential is an electric potential, such as was used in discussion of resistors and capacitors, then the electric field E is given by

(4.2)
$$E = -\text{gradient } V = -\frac{\partial V}{\partial n} \, \hat{n}.$$

This equation is independent of the choice of coordinates and depends only on the spacing of the equipotentials. This relation is therefore an *invariant* property of the field.

The gradient vector, $(\partial V/\partial n)\hat{n}$, can be written as the vector sum of its components in an arbitrary coordinate system. If x, y, z coordinates are chosen,

(4.3)
$$\text{gradient } V = \frac{\partial V}{\partial n} \, \hat{n} = \frac{\partial V}{\partial x} \, \hat{x} + \frac{\partial V}{\partial y} \, \hat{y} + \frac{\partial V}{\partial z} \, \hat{z}.$$

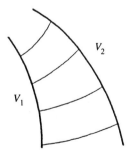

Fig. 4.2 Slices and the potential gradient.

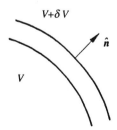

Fig. 4.3 Local gradient.

The gradient is always the maximum slope because it is perpendicular to the equipotentials.

By definition of the gradient

(4.4)
$$\int_1^2 \frac{\partial V}{\partial n} \, dn = V_2 - V_1.$$

In terms of geographical maps, the potential corresponds to a height above sea level, which is a measure of the potential energy in a gravitational field. In terms of electrical potential difference, it is the difference in the potential energy of a unit electrical charge. This is the work done in moving such a charge from a lower to a higher potential. This work is independent of the path taken. Consider this with respect to Fig. 4.4, which shows a path from a point 1 with a potential V_1, to a point 2 with a potential V_2. The force on the unit charge is F, uphill, and F is the direction of the vector $\hat{n}$ at the place where the path is in the direction dl. The work is done by the component of the force along the path, so that

(4.5)
$$\int_1^2 F \cos \theta \, dl = \int_1^2 F \cdot dl,$$

where the dot product (the scalar product) of two vectors is given by the product of their magnitudes and the cosine of the angle between them.

In terms of the gradient in an x, y, z coordinate system

(4.6)
$$\int_1^2 (\text{gradient } V) \cdot dl = \int_1^2 \left(\frac{\partial V}{\partial x} \hat{x} + \frac{\partial V}{\partial y} \hat{y} + \frac{\partial V}{\partial z} \hat{z} \right) \cdot (dx\hat{x} + dy\hat{y} + dz\hat{z}).$$

Now

(4.7)
$$\hat{x} \cdot \hat{x} = \hat{y} \cdot \hat{y} = \hat{z} \cdot \hat{z} = 1$$

and

(4.8)
$$\hat{x} \cdot \hat{y} = \hat{y} \cdot \hat{z} = \hat{z} \cdot \hat{x} = 0$$

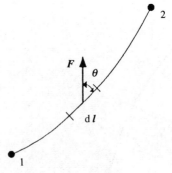

Fig. 4.4 Moving a charge between two points.

by definition of the dot product. Hence

(4.9)
$$\int_1^2 (\text{gradient } V) \cdot d\boldsymbol{l} = \int_1^2 \left(\frac{\partial V}{\partial x} dx + \frac{\partial V}{\partial y} dy + \frac{\partial V}{\partial z} dz \right) = V_2 - V_1.$$

As expected from the consideration of work done in a field of equipotentials, the integral in eqn (4.9) is independent of the choice of coordinates and depends only on the potentials at the ends of the path.

In terms of E, the electric field strength, which is downhill,

(4.10)
$$\int_1^2 -E \cdot d\boldsymbol{l} = V_2 - V_1.$$

For the closed path,

(4.11)
$$\oint (\text{gradient } V) \cdot d\boldsymbol{l} = 0.$$

No work is done in moving a charge in a path which begins and ends at the same potential. Such a field of potential energy is said to be conservative, because it conserves energy in a cyclic process.

For convenience, the gradient operator is often written gradient $V \equiv \text{grad } V \equiv \nabla V$. In x, y, z coordinates

(4.12)
$$\text{gradient } V = \nabla V = \left(\frac{\partial}{\partial x} \hat{x} + \frac{\partial}{\partial y} \hat{y} + \frac{\partial}{\partial z} \hat{z} \right) V.$$

The expression in brackets isolates the operator, and this is useful for mathematical manipulation. The operator differentiates and converts the scalar V into the vector ∇V. But of course the operator by itself has no physical significance. It needs to operate on a potential. The operator ∇ is often written as *del* or *nabla*, but it is wise to keep in mind the physical meaning implied by the word gradient.

4.3 The flux vector

In developing the idea of tubes and slices it was noticed that a thin tube has direction, because it can be made up of locally straight pieces. Hence the flux in such a tube has direction as well as magnitude. In the limit, such a flux is the product of the flux density and a small (infinitesimal) area across which the flux flows. A small area can be considered as flat, and it can be associated with the normal direction, which was met in the consideration of the gradient field.

Consider the flux of current such as that shown in Fig. 4.5, then

(4.13)
$$\delta I = J_n \, \delta S,$$

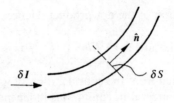

Fig. 4.5 A flux of current.

or in terms of vectors

(4.14)
$$\delta I = \boldsymbol{J} \cdot \delta \boldsymbol{S}.$$

Notice that the dot product converts the vectors $\boldsymbol{J}$ and $\delta \boldsymbol{S}$ into the scalar δI. A current is given by the number of amperes, which is a scalar, but the current density, in A m^{-2}, is a vector in the direction of the thin tube in which it flows. It is preferable to speak of thin tubes rather than streamlines, because to think of current flowing along a line implies an infinite current density, which is physically impossible.

If electric flux is considered, instead of the current, then

(4.15)
$$\delta \Psi = \boldsymbol{D} \cdot \delta \boldsymbol{S},$$

and for magnetic flux

(4.16)
$$\delta \Phi = \boldsymbol{B} \cdot \delta \boldsymbol{S}.$$

Remember that the idea of flux in electric and magnetic fields is connected with the inverse-square law, which describes the flow of an incompressible fluid. In electrostatics, the electric charge can be regarded as the source of the electric flux, so that

(4.17)
$$\Psi = \oint \boldsymbol{D} \cdot \mathrm{d}\boldsymbol{S} = \sum Q,$$

where the integral is to be evaluated over a surface enclosing the charges, Q.

The vector notation, which deals with a local tube of flux, can now be used to find an expression for the outflow from a small volume. First, the charges, Q, are converted into a distribution of charge density, ρ. Then

(4.18)
$$\sum Q = \int \rho \, \mathrm{d}v.$$

Then consider a small volume, so that

(4.19)
$$\int \rho \, \mathrm{d}v = \rho \delta v,$$

where the charge density, ρ, has been assumed to be constant throughout the small volume. The outflow from δv is given by

(4.20)
$$\oint \boldsymbol{D} \cdot \mathrm{d}\boldsymbol{S} = \rho \delta v.$$

Since ρ is finite, $(1/\delta v) \oint \boldsymbol{D} \cdot \mathrm{d}\boldsymbol{S}$ is also finite, and this is true as $\delta v \rightarrow 0$. Hence

(4.21)
$$\lim_{\delta v \to 0} \frac{1}{\delta v} \oint \boldsymbol{D} \cdot \mathrm{d}\boldsymbol{S} = \rho.$$

The left-hand side of this equation is the outflow per unit volume from a small volume. This equation is independent of the choice of coordinates, since ρ is just a number, and it therefore describes another invariant property of a vector field. For this reason, the left-hand side is given a distinctive name: the divergence of the field. We write

(4.22)
$$\text{divergence } \boldsymbol{D} \equiv \lim_{\delta v \to 0} \frac{1}{\delta v} \oint \boldsymbol{D} \cdot \mathrm{d}\boldsymbol{S},$$

which is equivalent to

(4.23)
$$\int \text{div } \boldsymbol{D} \, \mathrm{d}v = \int \boldsymbol{D} \cdot \mathrm{d}\boldsymbol{S}$$

for an arbitrary volume, v, enclosed by a surface, S.

Let us apply this to an x, y, z coordinate system. The infinitesimal volume, δv, in these coordinates is a small brick with sides δx, δy, and δz. This is shown in Fig. 4.6. Consider the outflow of the electric flux from such a volume, illustrated in Fig. 4.7, where only the x-component of $\boldsymbol{D}$ is shown. The outflow of D_x is given by

(4.24)
$$\left(D_x + \frac{\partial D_x}{\partial x} \delta x \right) \delta y \delta z - D_x \delta y \delta z = \frac{\partial D_x}{\partial x} \delta x \delta y \delta z.$$

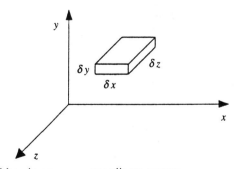

Fig. 4.6 A small box in an x, y, z coordinate system.

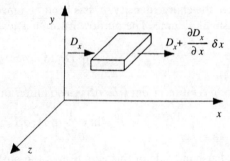

Fig. 4.7 The divergence of a vector field.

Hence the outflow of D_x, D_y, and D_z is given by

(4.25)
$$\oint \boldsymbol{D} \cdot d\boldsymbol{S} = \left(\frac{\partial D_x}{\partial x} + \frac{\partial D_y}{\partial y} + \frac{\partial D_z}{\partial z}\right) \delta x \delta y \delta z.$$

Hence

(4.26)
$$\text{div } \boldsymbol{D} = \frac{\partial D_x}{\partial x} + \frac{\partial D_y}{\partial y} + \frac{\partial D_z}{\partial z}.$$

There is some similarity between this and grad V given by

(4.27)
$$\text{grad } V = \left(\frac{\partial}{\partial x}\hat{\boldsymbol{x}} + \frac{\partial}{\partial y}\hat{\boldsymbol{y}} + \frac{\partial}{\partial z}\hat{\boldsymbol{z}}\right) V = \nabla V.$$

The divergence of a vector can be written as a scalar product of the vector operator and the vector, that is,

(4.28)
$$\text{div } \boldsymbol{D} = \left(\frac{\partial}{\partial x}\hat{\boldsymbol{x}} + \frac{\partial}{\partial y}\hat{\boldsymbol{y}} + \frac{\partial}{\partial z}\hat{\boldsymbol{z}}\right) \cdot (D_x\hat{\boldsymbol{x}} + D_y\hat{\boldsymbol{y}} + D_z\hat{\boldsymbol{z}}) = \nabla \cdot \boldsymbol{D}.$$

The term divergence is often abbreviated to div and

(4.29)
$$\text{div } \boldsymbol{D} \equiv \nabla \cdot \boldsymbol{D}.$$

Notice that the divergence of the vector $\boldsymbol{D}$ gives the charge density, ρ, which is a scalar. The divergence is a vector operator which converts a vector into a scalar, whereas the gradient is a vector operator which converts a scalar into a vector.

The use of the same operator, ∇, in the gradient and the divergence operations is however restricted to Cartesian coordinates, (x, y, z). We shall illustrate this by using cylindrical coordinates, (R, ϕ, z). For these coordinates, the gradient is given by the three components, $(\partial V/\partial R)\hat{\boldsymbol{R}}$,

$(1/R)\,(\partial V/\partial\phi)\hat{\phi}$, and $(\partial V/\partial z)\hat{z}$. The vector operator could therefore be written

$$V \equiv \left(\frac{\partial}{\partial R}\,\hat{R} + \frac{1}{R}\frac{\partial}{\partial\phi}\,\hat{\phi} + \frac{\partial}{\partial z}\,\hat{z} \right).$$

(4.30)

Notice that in the second term the length $R\delta\phi$ is in the denominator; this is because the gradient is the difference in potential divided by the length, so it would be incorrect to write this term as $(\partial/\partial\phi)\hat{\phi}$. Next, consider the divergence of the vector D in R, ϕ, z coordinates. The small volume is $\delta R R\delta\phi\delta z$, which is illustrated in Fig. 4.8. Notice that the curved length increases with the radius. For the component D_R, the outflow is

(4.31)

$$\left(D_R + \frac{\partial D_R}{\partial R}\,\delta R \right)(R+\delta R)\delta\phi\delta z - D_R R\delta\phi\delta z = D_R\delta R\delta\phi\delta z + \frac{\partial D_R}{\partial R}\,R\delta R\delta\phi\delta z,$$

where the term $(\partial D_R/\partial R)\delta R\delta\phi\delta z$ can be neglected. The total outflow is therefore

$$D_R\delta R\delta\phi\delta z + \frac{\partial D_R}{\partial R}\,R\delta R\delta\phi\delta z + \frac{\partial D_\phi}{\partial\phi}\,\delta\phi\delta R\delta z + \frac{\partial D_z}{\partial z}\,\delta z\delta R R\delta\phi$$

(4.32)

$$= \delta R R\delta\phi\delta z \left(\frac{D_R}{R} + \frac{\partial D_R}{\partial R} + \frac{1}{R}\frac{\partial D_\phi}{\partial\phi} + \frac{\partial D_z}{\partial z} \right).$$

Hence, dividing by the small volume,

(4.33)

$$\mathrm{div}\,D \equiv \frac{D_R}{R} + \frac{\partial D_R}{\partial R} + \frac{1}{R}\frac{\partial D_\phi}{\partial\phi} + \frac{\partial D_z}{\partial z}.$$

Compare this with the gradient operator in eqn (4.30) and notice that there is an additional term, D_R/R. Hence we cannot write $\mathrm{div}\,D = \nabla \cdot D$, where ∇ is the vector operator derived for the gradient. Indeed, eqn (4.32) does not allow separation of the operator from the vector on which it acts. Hence $\nabla \cdot D$ should be read as the divergence of D rather than del dot D. To obtain the divergence for curvilinear coordinates one must always go back to the definition of the outflow divided by the volume.

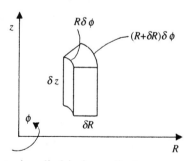

Fig. 4.8 A small volume in cylindrical coordinates.

Notice also that the divergence of the flux vector is equal to the local volume density of the charge. Hence, for electric fields, div $D = \rho$ in regions of charge density, and div $D = 0$ where there is no local charge density. Div $D = 0$ does not imply that $D = 0$ because the local D may be due to charges elsewhere. For steady electric currents div $J = 0$ everywhere, and for magnetic fields div $B = 0$ everywhere. The tubes of steady current and of magnetic flux are closed on themselves, whereas tubes of an electric field can start and end on surfaces carrying free charge. Fields of closed tubes of flux are called solenoidal fields because they behave like the magnetic fields of solenoids.

4.4 Laplace's and Poisson's equations

We have discussed three physical phenomena: steady currents, electrostatic capacitances, and magnetostatic interactions between magnets and steady currents. All of these can be described in terms of incompressible fluids flowing in tubes which are intersected at right angles by equipotential surfaces. Electrostatic fields are conservative and so are the magnetic fields of dipoles. On the other hand, the electric fields of the steady currents discussed in Chapter 1, and the magnetic fields of such currents, are solenoidal. This means that they are nonconservative. In steady-current flows in resistors, energy is converted into heat; and in the magnetic field of steady currents there is the magnetomotive force, $\oint H \cdot dl = I$, so that work is done if a magnetic pole is moved in a closed path linking a current.

In nonconservative fields it is not possible to map an entire system in equipotentials, but it is possible to do so for part of a system. For example, a steady current in a resistor can be divided into slices of potential, but the battery or generator driving the current must be excluded. Similarly, the magnetic field of a steady current can be treated as a potential field as long as there is no linkage with a current. In effect, the current is then replaced by magnetic shells, but the region occupied by the shells must be excluded.

Now consider the following mathematical description of a potential field taking electrostatic fields as an example. The field strength, as a force per unit charge, is given as the negative gradient of the potential, that is,

$$E = -\operatorname{grad} V.$$ (4.34)

If there is no local free charge density then

$$\operatorname{div} D = 0.$$ (4.35)

Also, D and E are connected by the *constitutive equation*

$$D = \varepsilon E.$$ (4.36)

Hence

(4.37) $$\text{div } \varepsilon \boldsymbol{E} = -\text{div}(\varepsilon \text{ grad } V) = 0.$$

Now $\varepsilon = \varepsilon_r \varepsilon_0$, where ε_0 is a constant throughout space, and ε_r depends on the presence of polarizable material. Assume that this material is homogeneous, so that ε_r is also a constant. Then

(4.38) $$\varepsilon \text{div}(\text{grad } V) = 0$$

and

(4.39) $$\text{div}(\text{grad } V) = 0.$$

In terms of the vector operator ∇ this can be written

(4.40) $$\nabla \cdot \nabla V = \nabla^2 V = 0.$$

It is easy to verify that for x, y, z coordinates

(4.41) $$\nabla^2 \equiv \frac{\partial^2}{\partial x^2} + \frac{\partial^2}{\partial y^2} + \frac{\partial^2}{\partial z^2}.$$

For curvilinear coordinates there will be more complicated expressions which must be obtained from the definitions of the gradient and the divergence and from the identity $\nabla^2 \equiv \text{div}(\text{grad})$. With this proviso $\nabla^2 V = 0$ can be treated as being independent of the coordinates and as being a property of a homogeneous conservative field. It is called Laplace's equation, and problems to which it applies are called Laplacian problems. The operator ∇^2 is called the Laplacian operator.

Any Laplacian problem can be solved by solving Laplace's equation, but this is not as easy as it sounds. The integration of this partial differential equation introduces constants of integration representing the physical and geometrical arrangement which gives rise to the potential distribution. There is, therefore, an infinity of possible solutions for Laplacian problems. The physical content of the equation is merely that there is a conservative field in a certain region, and that there are no free charges in that region. The second condition can be relaxed by putting

(4.42) $$\text{div } \boldsymbol{D} = \rho.$$

Then

(4.43) $$\nabla^2 V = -\frac{\rho}{\varepsilon},$$

this is called Poisson's equation. Solutions of this equation generally proceed via two steps. First, Laplace's equation is solved, and then the effect of the free-charge distribution is added. This procedure is analogous to finding the complementary function and particular integral of an ordinary differential equation. Of course this means that ρ must be known.

If not only the local charge density is known, but the entire charge distribution of the system, an expression for the potential can be written from the inverse-square law

$$(4.44) \qquad V = \sum \frac{Q}{4\pi\varepsilon r} = \int \frac{\rho}{4\pi\varepsilon r} \, dv.$$

In this expression, the distance r is the distance of the charge density, ρ, to the point at which V is to be calculated. (It is not the distance from the origin of coordinates, which can be arbitrary.) In a particular coordinate system, therefore, a distinction must be made between the *source* coordinates of ρ and the *field* coordinates of V. Notice that the differentiations in Laplace's and Poisson's equations are carried out on the field coordinates and that the integration of eqn (4.43) is carried out on the source coordinates. These operations are independent of each other. It is usual to distinguish the source coordinates by using primes, for example, x', y', z'. Then the distance r is

$$(4.45) \qquad r = [(x-x')^2 + (y-y')^2 + (z-z')^2]^{1/2}.$$

The procedure can be tested by verifying that eqns (4.43) and (4.44) are consistent with each other.

$$(4.46) \qquad \nabla^2 V = \nabla^2 \int \frac{\rho}{4\pi\varepsilon r} \, dv' = \int \nabla^2 \left(\frac{\rho}{4\pi\varepsilon r} \right) dv' = \frac{1}{4\pi\varepsilon} \int \rho \nabla^2 \left(\frac{1}{r} \right) dv'.$$

Notice the prime on the source volume, v'. Notice, also, that the operators ∇^2 and $\int$ can be interchanged because they are independent. Consider

$$(4.47) \qquad \nabla^2 \left(\frac{1}{r} \right) = \nabla^2 \left\{ \frac{1}{[(x-x')^2 + (y-y')^2 + (z-z')^2]^{1/2}} \right\}.$$

We find that $\nabla^2(1/r) = 0$ unless the field and source points coincide. Then, the integral over a small sphere as $r \to 0$ must be considered

$$\int \rho \nabla^2 \left(\frac{1}{r} \right) dv' = \rho \int \operatorname{div} \operatorname{grad} \left(\frac{1}{r} \right) dv'$$

$$(4.48) \qquad = \rho \int \operatorname{grad} \left(\frac{1}{r} \right) \cdot dS' = -4\pi\rho.$$

Hence $\nabla^2 V = 0$ everywhere where r is finite, and $\nabla^2 V = -\rho/\varepsilon$ at the point where there is a local charge ρ. In eqn (4.48), we have used the fact that ρ is constant in the small spherical volume and that the volume integration of the divergence is equal to the surface integration of the vector.

4.5 Polar and axial vectors

We have treated a small area δS as a vector in a direction normal to its locally flat surface. This is very convenient in the consideration of flux tubes, but it may hide the essential difference between an element of length and an element of area. Consider this in terms of coordinates: an element of length can be typified by δx, and an element of area by $\delta x \delta y$. If these elements are regarded as vectors, we have δx and it can be multiplied by δy to give δS. The scalar multiplication of vectors has already been dealt with, and a dot was as a symbol. A symbol is now needed for the vector multiplication of vectors. We write

(4.49)
$$\delta S = \delta x \times \delta y$$

and call this the *cross* or *wedge* multiplication. Clearly, δS must be perpendicular to both δx and δy, because its direction is normal to the area. A convention is needed to define the positive direction of the normal, and this is done by using right-handed axes; that is, δS is in the z-direction.

More generally

(4.50)
$$C = A \times B,$$

and the magnitude is given by

(4.51)
$$C = AB \sin \theta,$$

where θ is the angle between A and B, as in Fig. 4.9.

In describing a field, vectors associated with a length have been used – such as potential gradients or the magnetic field vector, H, which is associated with a length in the m.m.f. integral $\int H \cdot dl$. Flux vectors such as J, D, and B have also been used, and they are associated with an area. Physically, there is a difference between a potential difference (or an m.m.f.) and a flux. Vectors associated with a length are called *polar vectors* and vectors associated with an area are called *axial vectors*. These vectors are connected by the constitutive equations

(4.52)
$$J = \sigma E,$$

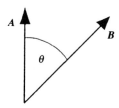

Fig. 4.9 Cross multiplication.

(4.53)
$$D = \varepsilon E,$$

(4.54)
$$B = \mu H.$$

It can now be seen that ρ, ε, and μ transform the polar vectors E and H into the axial vectors J, D, and B. The question arises as to whether either the polar or the axial vectors can be regarded as the 'cause' and the other type as the effect. One or other type of vector could then be treated as fundamental.

It is better to remember that we are dealing with the energy densities $\frac{1}{2}EJ \times$ time, $\frac{1}{2}ED$, or $\frac{1}{2}HB$. It is the energy which is fundamental, and the polar and axial vectors are aspects of the energy distribution. In terms of tubes and slices, note that both methods need a volume for calculation of the energy. The tubes have to have a length as well as an area and the slices have to have an area as well as a distance between the equipotentials. Axial and polar vectors always occur together, just as area and length occur together to define a volume.

4.6 Vortex fields

Conservative fields are associated with potential energies and they cannot produce a cyclic or vortex flow. In Chapter 3 it was observed that the magnetic energy of electric currents is analogous to kinetic energy, and this type of energy can be associated with a vortex flow. An example of this is given by the relationship

(4.55)
$$\oint H \cdot dl = I$$

for currents, which describes a magnetomotive force acting around a closed loop. For magnets the following conservative relationship should apply instead

(4.56)
$$\oint H \cdot dl = 0.$$

Let us now look at these equations in the local differential form. We will proceed as for Gauss's law in Section 4.3, where an arbitrarily large volume was replaced by an infinitesimal volume. Here, an arbitrary large loop needs to be replaced by a small loop. In order to keep the current density finite we write

(4.57)
$$I = \int J \cdot dS,$$

where S is an area bounded by a loop, and

(4.58)
$$\oint H \cdot dl = \int J \cdot dS.$$

For a small loop,

(4.59) $$\oint \boldsymbol{H} \cdot d\boldsymbol{l} = \boldsymbol{J} \cdot \delta \boldsymbol{S} = J_n \delta S.$$

Hence

(4.60) $$\lim_{\delta S \to 0} \frac{1}{\delta S} \oint \boldsymbol{H} \cdot d\boldsymbol{l} = J_n = \hat{\boldsymbol{n}} \cdot \boldsymbol{J},$$

(4.61) $$\lim_{\delta S \to 0} \frac{\hat{\boldsymbol{n}}}{\delta S} \oint \boldsymbol{H} \cdot d\boldsymbol{l} = \boldsymbol{J}.$$

Since $\boldsymbol{J}$ is independent of the coordinates, the left-hand side is also an invariant of the field. It is a measure of the local vorticity and is called curl $\boldsymbol{H}$. It is a vector perpendicular to the loop. Its association with an area shows it to be an axial vector. By definition

(4.62) $$\int (\mathrm{curl}\ \boldsymbol{H}) \cdot d\boldsymbol{S} = \oint \boldsymbol{H} \cdot d\boldsymbol{l}.$$

This is known as Stokes' theorem and it is reminiscent of Ampère's device of turning a large current loop into a sum of the small loops in an area. The area S is bounded by the perimeter l, but it does not need to be flat except locally.

Let us now find an expression for curl $\boldsymbol{H}$ in x, y, z coordinates. Figure 4.10 shows a small area in the xy-plane; consider the line integral of $\boldsymbol{H}$ around its perimeter, starting at the bottom left-hand corner and proceeding anti-clockwise.

(4.63) $$H_x \delta x + \left(H_y + \frac{\partial H_y}{\partial x} \delta x \right) \delta y - \left(H_x + \frac{\partial H_x}{\partial y} \delta y \right) \delta x - H_y \delta y$$

$$= \left(\frac{\partial H_y}{\partial x} - \frac{\partial H_x}{\partial y} \right) \delta x\ \delta y.$$

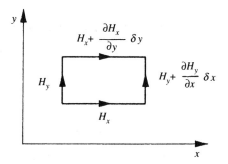

Fig. 4.10 A derivation of curl.

Hence

$$(4.64) \qquad (\text{curl } \boldsymbol{H})_z = \frac{1}{\delta x \delta y} \oint \boldsymbol{H} \cdot d\boldsymbol{l} = \frac{\partial H_y}{\partial x} - \frac{\partial H_x}{\partial y} = J_z.$$

The other two components of curl $\boldsymbol{H}$ can be obtained by considering the yz-plane and the zx-plane. A concise way of writing curl $\boldsymbol{H}$ is in the form of the following determinant

$$(4.65) \qquad \text{curl } \boldsymbol{H} = \begin{vmatrix} \hat{\boldsymbol{x}} & \hat{\boldsymbol{y}} & \hat{\boldsymbol{z}} \\ \dfrac{\partial}{\partial x} & \dfrac{\partial}{\partial y} & \dfrac{\partial}{\partial z} \\ H_x & H_y & H_z \end{vmatrix}$$

The vector operator ∇ can also be used to write

$$(4.66) \qquad \text{curl } \boldsymbol{H} \equiv \nabla \times \boldsymbol{H}.$$

Our previous experience with curvilinear coordinates will, however, warn us that the separation of ∇ and $\boldsymbol{H}$ in eqn (4.66) is possible only in x, y, z coordinates. In general, the expression $\nabla \times \boldsymbol{H}$ must be understood as curl $\boldsymbol{H}$, and this must be derived from its definition as the line integral for a small area divided by that area. The right-handed-screw convention is illustrated in Fig. 4.11, which shows a 'curl meter'.

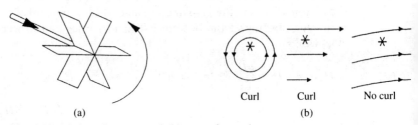

(a)   Curl   Curl   No curl   (b)

Fig. 4.11 (a) A curl meter, and (b) a test for curl.

4.7 The independence of conservative and vortex fields

A conservative field can be expressed in terms of the gradient of a scalar potential. For example,

$$(4.67) \qquad \boldsymbol{E} = -\nabla V.$$

Hence

$$(4.68) \qquad \nabla \times \boldsymbol{E} = -\nabla \times \nabla V = 0.$$

It is clear that the cross product of a vector with itself is zero, because the vectors are parallel to each other so that the sine of the angle between them is zero. The vector operator ∇ is used in eqn (4.68), and it is easy to show that $\nabla \times \nabla = 0$ in x, y, z coordinates. But, if an invariant vector expression is zero in one set of coordinates, it must be zero for all sets of coordinates, so we do not have to investigate various types of curvilinear coordinates. Hence the curl of a conservative field is zero. This of course follows from the definition of curl as a line integral. A gradient field has no vorticity.

Next consider a curl field such as

$$\text{curl } \boldsymbol{H} = \nabla \times \boldsymbol{H} = \boldsymbol{J}. \tag{4.69}$$

This field has no divergence, since

$$\nabla \cdot \nabla \times \boldsymbol{H} = 0. \tag{4.70}$$

This follows from the fact that a scalar and vector product of the same vector is zero. A vector cannot have a component perpendicular to itself. This is true for this vector operator in x, y, z coordinates, and since it is an invariant vector relationship it is true for all types of coordinates. Hence a curl field has no divergence. For example the relationship

$$\text{curl } \boldsymbol{H} = \boldsymbol{J} \tag{4.71}$$

implies that

$$\text{div } \boldsymbol{J} = 0. \tag{4.72}$$

There are therefore two types of 'sources' for vector fields: polar vectors like $\boldsymbol{H}$ can have *curl* sources and axial vectors like $\boldsymbol{D}$ can have *divergence* sources. Sometimes, both types of source occur together, for example a magnetic field may be modelled in terms of electric currents and magnetic poles. The currents give curl $\boldsymbol{H}$, and the poles give div $\boldsymbol{B}$. Similarly, the electric field of nonconservative sources, such as batteries and generators, can be modelled in terms of curl $\boldsymbol{E}$, which can be accompanied by the electric charge given by div $\boldsymbol{D}$. Thus both the slices and the tubes can be associated with their own type of source. In general, the energy, which is the product of the polar and axial vectors, can depend on both types of sources.

However, a field of divergence sources only is always conservative, and a field of curl sources only is always solenoidal. Figure 4.12 illustrates the various types of field.

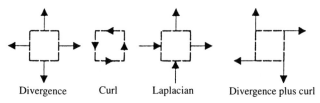

Fig. 4.12 Types of fields.

4.8 The magnetic vector potential

The idea of potential was originally derived from the potential energy associated with conservative fields, such as the gravitational field of attracting masses, the electrostatic field of stationary charges, and the magnetostatic field of stationary poles or dipoles. For example, if in a region curl $E = 0$, we can write $E = -\text{grad } V$. A vortex field, such as the magnetic field curl $H = J$, is associated with the kinetic energy and it cannot be described in terms of a scalar potential. Nevertheless, a solenoidal field for which div $B = 0$ can be dealt with in terms of a *vector potential* as follows. Since

(4.73)
$$\text{div curl } A = 0$$

we can put

(4.74)
$$B = \text{curl } A.$$

This is true everywhere because there are no free magnetic poles. (Although it is sometimes convenient to model magnetic fields in terms of a fictitious polarity.)

In magnetostatics, curl $H = J$, so that

(4.75)
$$\text{curl}\left(\frac{1}{\mu}\text{curl } A\right) = J.$$

If μ is constant this gives

(4.76)
$$\text{curl curl } A = \mu J.$$

A useful vector identity, which is left as an exercise for the reader, is

(4.77)
$$A \times (B \times C) = B(A \cdot C) - C(A \cdot B).$$

To remember this, read the right-hand side as $bac - cab$. Hence

(4.78)
$$\nabla \times \nabla \times A = \nabla(\nabla \cdot A) - \nabla^2 A.$$

A little caution is needed in the use of eqn (4.78), because $\nabla \cdot \nabla(A)$ cannot be read as div grad A, since one cannot take the gradient of a vector. However, in x, y, z coordinates the operator $\nabla \cdot \nabla \equiv \nabla^2$ can be applied to the three components of A. For curvilinear coordinates, $\nabla \cdot \nabla(A)$ needs to be written in full, as derived from the eqn (4.78).

We can put

(4.79)
$$\text{div } A = 0$$

because A has so far been defined only in terms of its curl sources, B. Hence, from eqn (4.76)

(4.80)
$$\text{curl curl } A = -\nabla^2 A = \mu J$$

and

(4.81)
$$\nabla^2 A = -\mu J.$$

This gives three equations

(4.82)
$$\nabla^2 A_x = -\mu J_x,$$

(4.83)
$$\nabla^2 A_y = -\mu J_y,$$

(4.84)
$$\nabla^2 A_z = -\mu J_z.$$

Comparison with eqn (4.48) gives the integral

(4.85)
$$A_x = \int \frac{\mu J_x}{4\pi r} \, dv',$$

and there are similar terms in y and z. Notice that A_x arises only from J_x, and similarly A_y arises from J_y, and A_z arises from J_z. Mathematically, there is a close analogy between the scalar potential V and the vector A. For this reason, A is called the (magnetic) vector potential. This name is unfortunate from a physical point of view because potential energy does not have direction and cannot be a vector. James Clerk Maxwell called A the *electrokinetic momentum*. This refers to the fact that the magnetic field describes the kinetic energy of current and that the vector nature of A shows it to be a momentum rather than an energy. We shall find the idea of momentum very appropriate when we come to deal with time-varying fields.

4.9 The uniqueness of vector fields

Differential equations like Laplace's and Poisson's equations and the equation for the vector potential describe the behaviour of fields in a local neighbourhood. This is insufficient to determine the fields uniquely, since there may be undefined sources at other places. On the other hand, the integral expressions for the potentials do give a unique value to the fields, but only if all the sources are specified. In general these sources are known only in a limited region and it is important to determine what conditions are needed on the surface enclosing the region to make the internal field unique.

Consider first a conservative system. The divergence theorem gives

(4.86)
$$\int \text{div} \, F \, dv = \oint F \cdot dS,$$

where the surface S encloses the volume v. Let

(4.87)
$$F = \phi \nabla \phi.$$

Then

(4.88)
$$\nabla \cdot \boldsymbol{F} = (\nabla \phi)^2 + \phi \nabla^2 \phi.$$

Hence

(4.89)
$$\int [(\nabla \phi)^2 + \phi \nabla^2 \phi] \, dv = \oint \phi \nabla \phi \cdot d\boldsymbol{S}.$$

Put $\phi = \phi_1 - \phi_2$, where ϕ_1 and ϕ_2 are two separate solutions for the potential of the system. If the internal sources are known, then

(4.90)
$$\nabla^2 \phi_1 = \nabla^2 \phi_2 = -\frac{\rho}{\varepsilon}$$

and therefore

(4.91)
$$\nabla^2 \phi = \nabla^2 \phi_1 - \nabla^2 \phi_2 = 0.$$

Hence eqn (4.89) gives

(4.92)
$$\int (\nabla(\phi_1 - \phi_2))^2 \, dv = \oint (\phi_1 - \phi_2) \nabla(\phi_1 - \phi_2) \cdot d\boldsymbol{S}.$$

If the surface S is made up of pieces on which either $\phi_1 = \phi_2$ or $\partial \phi_1 / \partial n = \partial \phi_2 / \partial n$, the surface integral is zero. Then

(4.93)
$$\int (\nabla(\phi_1 - \phi_2))^2 \, dv = 0.$$

Since the integrand is a squared quantity it is either positive or zero, and since its integral is zero it must itself be zero. Therefore,

(4.94)
$$\nabla(\phi_1 - \phi_2) = 0.$$

Hence the potential gradient is unique, and the potential differs at most by an arbitrary constant. Hence the field is unique. There can be only one solution. Uniqueness is assured if the internal charge, ρ is known and if on the enclosing surface the potential or the normal component of the field are specified. These surface conditions therefore represent all the external sources. They do not specify those external sources uniquely, but they represent their effect inside the volume.

Next, consider a vortex field described by a vector potential. For a vector $A \times \text{curl } A$, the divergence theorem gives

(4.95)
$$\int \text{div}(A \times \text{curl } A) \, dv = \oint (A \times \text{curl } A) \cdot d\boldsymbol{S}.$$

Now

(4.96)
$$\text{div}(A \times \text{curl } A) = (\text{curl } A)^2 - A \cdot \text{curl curl } A.$$

If $A = A_1 - A_2$, where A_1 and A_2 are two separate solutions for the vector potential, and if the internal current sources are known, then

(4.97)
$$\text{curl curl } A_1 = \text{curl curl } A_2 = \mu J$$

and

(4.98)
$$\text{curl curl } A = \text{curl curl } A_1 - \text{curl curl } A_2 = 0.$$

Hence

(4.99)
$$\int (\text{curl } A)^2 \, dv = \oint (A \times \text{curl } A) \cdot dS.$$

If the surface, S, is made up of pieces on which either the tangential components of $A_1 = A_2$ or the tangential components of $\text{curl } A_1 = \text{curl } A_2$, then the surface integral is zero. Then

(4.100)
$$\int [\text{curl}(A_1 - A_2)]^2 \, dv = 0.$$

Hence

(4.101)
$$\text{curl}(A_1 - A_2) = 0.$$

Hence the curl of the vector potential is unique and the vector potential differs at most by a gradient vector, and hence the field is unique.

It should be noted that this argument has omitted all reference to possible interaction between the potential energy of divergence sources and the kinetic energy of curl sources. Such interaction will be considered in the next chapter. Our argument so far applies to fields which have no time variation, that is, to the fields of stationary charges and magnets, and to stationary steady currents.

Exercises

4.1 Given

$$A = 3\hat{x} - 2\hat{y} + \hat{z}, \qquad B = 2\hat{x} - 4\hat{y} - 3\hat{z}, \qquad C = -\hat{x} + 2\hat{y} + 2\hat{z},$$

find the magnitudes of (a) C, (b) $A + B + C$, and (c) $2A - 3B - 5C$. (*Answer* (a) 3, (b) $4\sqrt{2}$, and (c) $\sqrt{30}$.)

4.2 What is the scalar product of any two diagonals through the body of a cube of side a? (*Answer* a^2.)

4.3 Find the work done in moving an object along a straight line from (3, 2, −1) to (2, −1, 4) in a force field given by $F = 4\hat{x} - 3\hat{y} + 2\hat{z}$. (*Answer* 15.)

4.4 Show that the area of a parallelogram of sides A and B is given by $S = A \times B$. What is meant by the direction of an area?

4.5 If

$$A = \hat{x} - 2\hat{y} - 3\hat{z}, \qquad B = 2\hat{x} + \hat{y} - \hat{z}, \qquad C = \hat{x} + 3\hat{y} - 2\hat{z},$$

find:

    (a) $|(A \times B) \times C|$,
    (b) $|A \times (B \times C)|$,
    (c) $A \cdot (B \times C)$,
    (d) $(A \times B) \cdot C$,
    (e) $(A \times B) \times (B \times C)$, and
    (f) $(A \times B)(B \cdot C)$.

(*Answer* (a) $5\sqrt{26}$, (b) $3\sqrt{10}$, (c) -20, (d) -20, (e) $-40\hat{x} - 20\hat{y} + 20\hat{z}$, and (f) $35\hat{x} - 35\hat{y} + 35\hat{z}$.)

4.6 Find the area of a parallelogram having diagonals $A = 3\hat{x} + \hat{y} - 2\hat{z}$ and $B = \hat{x} - 3\hat{y} + 4\hat{z}$.
(*Answer* $5\sqrt{3}$.)

4.7 Show that $A \times (B \times C) = B(A \cdot C) - C(A \cdot B)$.

4.8 Simplify $(A + B) \cdot (B + C) \times (C + A)$.
(*Answer* $2A \cdot B \times C$.)

4.9 Prove that $(A \times B) \cdot (C \times D) + (B \times C) \cdot (A \times D) + (C \times A) \cdot (B \times D) = 0$.

4.10 Find the shortest distance from $(6, -4, 4)$ to the line joining $(2, 1, 2)$ and $(3, -1, 4)$.
(*Answer* 3.)

4.11 Explain what is meant by the *gradient* of a scalar quantity. If $\phi(x, y, z) = 3x^2 y - y^3 z^2$, find $\nabla\phi$ (or grad ϕ) at the point $(1, -2, -1)$.
(*Answer* $\nabla\phi = -12\hat{x} - 9\hat{y} - 16\hat{z}$.)

4.12 Find $\nabla\phi$ when (a) $\phi = \ln|r|$ and (b) $\phi = 1/r$, where $r = x\hat{x} + y\hat{y} + z\hat{z}$ and $|r| = \sqrt{x^2 + y^2 + z^2}$.
(*Answer* (a) r/r^2, and (b) $-r/r^3$.)

4.13 Show that $\nabla\phi$ is a vector perpendicular to the surface $\phi(x, y, z) = c$ where c is a constant.

4.14 Explain what is meant by the *divergence* of a vector field. Discuss why an electric field, D, can have divergence, but a magnetic field, B, cannot have divergence.

4.15 Show that div $r = 3$.

4.16 If $A = x^2 z\hat{x} - 2y^3 z^2 \hat{y} + xy^2 z\hat{z}$, find $\nabla \cdot A$ (or div A) at the point $(1, -1, 1)$.
(*Answer* -3.)

4.17 Show that $\nabla \cdot \nabla\phi = \nabla^2\phi$, where ∇^2 denotes the Laplacian operator.

4.18 Prove that $\nabla^2(1/r) = 0$, except at $r = 0$, and thus that $\phi = 1/r$ is a solution of Laplace's equation.

4.19 Show that $\nabla \cdot (r/r^3) = 0$, except at $r = 0$.

4.20 A vector has components $A_R = k/R$, $A_\phi = 0$, and $A_z = 0$. Show that this vector can be derived from a scalar potential and find its value.
(*Answer* $V = -k \ln R$.)

4.21 Explain what is meant by the *curl* of a vector field. If $A = xz^3\hat{x} - 2x^2 yz\hat{y} + 2yz^4\hat{z}$, find $\nabla \times A$ (or curl A) at the point $(1, -1, 1)$.
(*Answer* $3\hat{y} + 4\hat{z}$.)

4.22 If $A = x^2 y\hat{x} - 2xz\hat{y} + 2yz\hat{z}$, find $\nabla \times \nabla \times A$.
(*Answer* $(2x+2)\hat{y}$.)

4.23 Derive the following relationships:

$$\nabla \times \nabla V = 0,$$
$$\nabla \cdot \nabla \times F = 0,$$
$$\nabla \times \nabla \times F = \nabla(\nabla \cdot F) - \nabla^2 F.$$

Electromagnetic induction

5.1 The importance of time

Chapter 4 was a digression in which a concise method of handling vector fields was acquired without the need for cumbersome and arbitrary coordinates. We can now treat vectors as single objects instead of having to deal with their projections. Moreover, invariant properties, such as the divergence and the curl, were found which give mathematical descriptions of the underlying physical processes in electromagnetic fields. However, it would be a mistake to regard a knowledge of vector algebra as anything more than a useful tool. In this we follow the teaching of James Clerk Maxwell whose pioneering work provided the first coherent account of electromagnetism. In his great *Treatise on Electricity and Magnetism* he wrote 'we avail ourselves of the labours of the mathematicians and retranslate their results from the language of the calculus into the language of dynamics, so that our words may call up the mental image, not of some algebraical process, but of some property of moving bodies'. Our object in this book is to sharpen the reader's physical insight. Such insight is far more important than mathematical fluency, although that is very useful. The reader who found the chapter on vectors rather difficult or the reader who decided to skip that chapter need not feel dismayed when he meets vector notation in this and the following chapters, because we shall always try to explain such notation in terms of physical processes. In this way we hope that readers will be won over and will go back to Chapter 4 to acquire the use of the mathematical methods described there.

Let us now return to an examination of the relationship between electric and magnetic fields. We recall that in Chapter 1 the flow of steady currents in resistors was associated with a potential difference. The potential difference can be thought of as driving the current, or the current can be thought of as producing a potential difference, these are two aspects of the phenomenon of the conversion of electrical energy into heat. The current can be subdivided into tubes and the potential difference into slices. The energy per unit time is associated with the volume of all the tubes or all the slices and this method provides numerical upper and lower bounds for the resistance. The gradient of the potential difference is the local electric field strength, which is related to the local current density by the conductivity of the material. This is the local form of Ohm's law and it is written as

$$\sigma E = J,$$

(5.1)

where E is a polar vector acting along a line and J is an axial vector acting across an area.

In Chapter 2 the same potential difference and electric field were met in the context of the potential energy of stationary charges. We found that the electric field was analogous to the flow of an incompressible fluid and that the energy could be obtained from the volume of the tubes of flux or of the slices of potential difference. The flux density and field strength are related by

$$\varepsilon E = D.$$

(5.2)

By comparing eqns (5.1) and (5.2), it can be seen that steady current flow is associated with electrostatic fields and it can be deduced that there is a charge distribution on the surface of resistors which provides the potential difference. However, electrostatic fields are conservative and they cannot provide the steady flow of energy which is required by Ohm's law. By itself, a capacitor connected to a resistor would become discharged and the current would stop flowing. There must also be nonconservative sources, such as batteries, which maintain the potential difference and the associated surface charges. There is also another important difference between eqns (5.1) and (5.2). In free space, $\sigma = 0$, but $\varepsilon = \varepsilon_0$ and is not zero. Hence current is confined to conductors, but electric flux can exist anywhere in space, so that there is electrical energy in free space.

In Chapter 3, there was a discussion of the connection between electricity and magnetism. Ampère's equivalence between current loops and magnets, and the forces between currents, between magnets, and between magnets and currents lead to the conclusion that electric currents are surrounded by a magnetic field which is related to the kinetic energy of the current. We found that the magnetic field could again be divided into tubes and slices, and that the flux density and magnetic field strength are related by

$$\mu H = B.$$

(5.3)

In free space $\mu = \mu_0$, so that the magnetic energy like its electric counterpart can exist in free space as well as in magnetic materials. Since ε is associated with the energy of stationary charges and μ with the energy of currents (which are moving charges), there is a link between ε and μ which can be given in terms of a velocity. This is

$$\varepsilon\mu = \frac{1}{v^2}$$

(5.4)

and in free space

(5.5)
$$\varepsilon\mu = \frac{1}{c^2},$$

where c is the velocity of light. This relationship was known before Maxwell's work as a relationship between electrostatic and magnetostatic energy, and it was used by Maxwell as a pointer towards the fact that light is an electromagnetic phenomenon. It was deduced from Ampère's equivalence between currents and magnets, and the value of c was found to be approximately the same as that which had been obtained by astronomical observations. Before Maxwell's work, eqn (5.5) was little more than a curious coincidence between two independent observations.

The existence of a fundamental velocity linking electric and magnetic effects points to the importance of time as a factor in electromagnetic interactions. These interactions cannot be understood in terms of spatial position alone; they need the additional specification of a time coordinate. The full implications of this took almost a century to unravel and led to enormous technological developments, not least in telecommunication systems. Time is also involved in the relationships between σ and ε, and σ and μ, because σ describes the rate of energy conversion, whereas ε and μ describe the energy.

5.2 Motional electromotive force and electromechanical energy conversion

Ampère's equivalence between currents and magnets already contained a time-dependence locally, although this was masked by treating steady-current systems as a whole. In order to look at the local effect the steady current must be replaced by a succession of moving charges, as illustrated in Fig. 5.1. Let the charges be spaced at a distance δl, and let them all move with a velocity v. This is, of course, a highly simplified picture of a current. There is a time interval given by

(5.6)
$$\delta t = \frac{\delta l}{v}$$

and the average current can be expressed as

(5.7)
$$I = \frac{Q}{\delta t}.$$

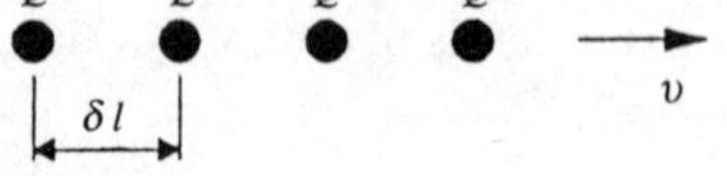

Fig. 5.1 Moving charges.

Hence

(5.8)
$$I \, \delta l = Qv.$$

A current element is equivalent to a charge multiplied by a velocity. It will be remembered that great care had to be taken in the treatment of current elements, because an isolated element will not conduct current. Similarly, we must remember that Q is not an isolated charge, but one charge in a steady succession of charges. Indeed, an isolated charge is just as impossible as an isolated current element. The tubes of flux which diverge from a positive charge must converge on a negative charge. Tubes of flux in a conservative field cannot be closed on themselves. Moreover the drift velocity, v, is a relative motion between positive and negative charges in a conductor.

However, the force on an element of a current circuit can be measured by providing flexible connections, and in principle it is possible to measure the force on a moving charge. From Chapter 3 (eqn (3.53))

(5.9)
$$F = I \, \delta l B_{\perp}$$

so that

(5.10)
$$F = Qv B_{\perp}.$$

In vector notation, these equations can be written

(5.11)
$$\boldsymbol{F} = I \, \delta \boldsymbol{l} \times \boldsymbol{B}$$

and

(5.12)
$$\boldsymbol{F} = Q\boldsymbol{v} \times \boldsymbol{B}.$$

But the electric field strength, E, was previously defined as a force per unit charge. Hence

(5.13)
$$\boldsymbol{E} = \boldsymbol{v} \times \boldsymbol{B}.$$

We have now shown that there are at least two types of force on electric charges. The first is the conservative force given by

(5.14)
$$\boldsymbol{E} = -\text{grad } V$$

which arises from stationary charges, and the second is a force on charges moving across magnetic fields. This force cannot be derived from the gradient of a potential, and it arises from changes of position with time. We shall see that the time-dependent force is nonconservative. Energy can be interchanged with charges moving in closed loops. Consider this with reference to Fig. 5.2, which shows a circuit linking a magnetic flux of constant density, $\boldsymbol{B}$, in space and time. A part of the circuit consists of a sliding conductor which moves at velocity v perpendicular to its length. The

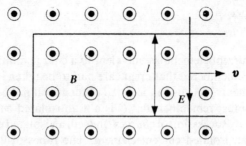

Fig. 5.2 A circuit linking a magnetic flux.

charges in the sliding conductor experience an electric field, $E = v \times B$, which is constant along the length l and is in the direction indicated by the arrow. Now consider the integration of E around the circuit. In vector notation this is given by

$$\oint E \cdot dl = (v \times B) \cdot l = vBl.$$

(5.15)

For electrostatic fields where l is taken in the direction of E

(5.16)
$$\oint E \cdot dl = 0.$$

It is important to note that the closed integral of eqn (5.15) would not be altered if in addition to the motional $E = v \times b$ there existed some electrostatic field $E = -\operatorname{grad} V$. Locally there should then be a total field, E, of

(5.17)
$$E = v \times B - \operatorname{grad} V,$$

but around the loop the effects of the electrostatic field would cancel out. Hence the loop integral isolates the nonconservative effect. For this reason it is called the *electromotive force* or, e.m.f. Notice also that the conservative and nonconservative electric field both have units of volts per metre $(\mathrm{V\ m^{-1}})$. There are therefore two kinds of voltage: electromotive force and potential difference.

The treatment in eqn (5.15) can be applied to a loop moving in a more general manner by writing

(5.18)
$$\oint E \cdot dl = \oint (v \times B) \cdot dl.$$

Here, every element δl has its own motion v across a magnetic flux of local density B.

The relationship between the motional electric field $E = v \times B$ and the

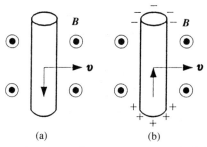

(a)   (b)

Fig. 5.3 An isolated conductor moving across a magnetic field.

electrostatic field $E = -\operatorname{grad} V$ can be clarified by considering an isolated conductor moving across a magnetic field as shown in Fig. 5.3. Figure 5.3(a) shows the $v \times B$ field, which will act on the charges in the conductor and will produce a surface charge distribution, as shown in Fig. 5.3(b) After an initial transient there will be no current and the two components of the electric field will cancel each other; that is,

(5.19)

$$v \times B = \operatorname{grad} V.$$

Now suppose that a stationary voltmeter is inserted into the circuit of Fig. 5.2, as shown in Fig. 5.4. The surface charges in Fig. 5.3(b) will be spread along the conductors connecting the moving conductor to the voltmeter and this will register the potential difference, which from eqn (5.19) will be equal to the e.m.f. When the motion stops, the $v \times B$ field disappears. There is then nothing to separate the charges, and the potential difference also disappears.

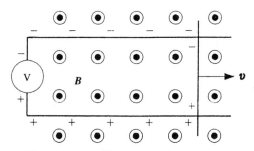

Fig. 5.4 A circuit with a voltmeter, V.

Now suppose that a stationary resistor is placed in parallel with the voltmeter, as in Fig. 5.5. A current will now circulate as shown. If the resistance of the connections is negligible compared to R, then by Ohm's law

(5.20)

$$\text{p.d.} = IR$$

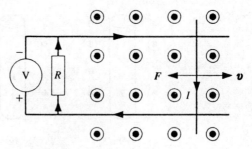

Fig. 5.5 A circuit with a resistor, R.

and therefore

(5.21)
$$\text{e.m.f.} = IR.$$

The power in the resistor is

(5.22)
$$I^2R = I(\text{p.d.}) = I(\text{e.m.f.}) = IvBl.$$

The moving conductor experiences a force

(5.23)
$$F = Il \times B,$$

where l is in the direction of the current flow and B is upwards out of the paper. Hence, F is (by the right-hand-screw rule) to the left. That means that the motion can be maintained only if an external force, $-F$ is applied in the direction of v. Hence mechanical power has to be supplied. This is given by

(5.24)
$$-F \cdot v = -I(l \times B) \cdot v = I(v \times B) \cdot l = IvBl.$$

Hence, as expected, energy is conserved and the mechanical input is equal to the heat generated in the resistor. There are in fact two energy-conversion processes. First, mechanical energy is converted into electrical energy, and secondly electrical energy is converted into heat. In the first conversion there is what is known as an electrical generator. The resistor could be replaced by other components which need electrical energy for their operation.

The operation illustrated by Fig. 5.5 is fully reversible if the resistor is replaced by a source of electrical energy like another generator or a battery. This is illustrated in Fig. 5.6. The current is now reversed. This reverses the force acting on the moving conductor, which now acts in the direction of v. Hence there is mechanical output and this output is again equal to the electrical input from the generator. The moving conductor now behaves as a motor. The potential difference applied to its ends causes the current to flow. This produces force and motion, and the motion produces an e.m.f.

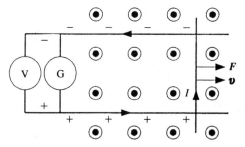

Fig. 5.6 A circuit with a generator, G.

opposing the potential difference. The process becomes steady when the e.m.f. field exactly cancels the potential difference field.

In practice, the moving conductor will have resistance and the potential difference for the motor will have to be slightly larger than the e.m.f. to allow for the Ohmic drop. If the resistance of the conductor is r, then

(5.25)
$$\text{p.d.} - \text{e.m.f.} = Ir$$

for the motor, and

(5.26)
$$\text{e.m.f.} - \text{p.d.} = Ir$$

for the generator.

It is interesting to consider how the force, F, acts on the current in the wire, when the current is analysed in terms of moving charges. Consider this with reference to Fig. 5.7, where the current is assumed to flow as in the motor shown in Fig. 5.6. The current is equivalent to a flow of charge Q moving with velocity u. Hence Q will be acted on by an electric field $E = u \times B$, as shown in Fig. 5.7. E will displace the moving charges horizontally, and a surface charge will appear producing an equal and opposite gradient field, $-E$. The force on the conductor will be due to the force on the surface charges, and it will be from left to right.

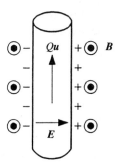

Fig. 5.7 Movement of positive charge.

This assumes a motion of positive charge Q upwards. Suppose, however, that the carrier charges are mobile negatively charged electrons. This situation is shown in Fig. 5.8. The electric field is reversed, but by its definition, E is the force on a positive charge. Hence E will move the electrons to the right as indicated. The force on the conductor will again be from left to right, but the potential difference between the sides of the conductor will have the opposite sign. This potential difference is known as the Hall voltage and its sign can be used to determine whether the carriers are electrons or positive holes, as might be the case in a semiconductor.

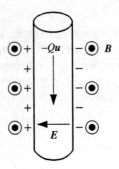

Fig. 5.8 Movement of negative charge.

5.3 The motion of charge in free space

The effect of motional electric force is not confined to the motion of charge in metallic conductors or semiconducting materials. It applies equally to the motion of charges in arcs and plasmas, where there are both moving electrons and moving positive ions, and also to vacuum discharges.

An interesting example is the cathode-ray tube illustrated in Fig. 5.9. Electrons are evaporated from a hot cathode and accelerated by a high-voltage anode, which has a hole through which some of the electrons pass. The electrons have a high velocity perpendicular to the hole, but because the hole has finite dimensions the electrons will also have a range of (smaller) transverse velocities.

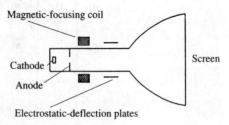

Fig. 5.9 A cathode-ray tube.

Now consider the effect of the magnetic coil, which produces a flux density **B** in the direction of the electron beam. As a result of their transverse velocities, v, the electrons will experience a force

(5.27)
$$F = evB$$

in a direction perpendicular to v and **B**. Since this force is perpendicular to v it will not alter the magnitude of v but only its direction. Moreover, the force itself will have a constant magnitude. Hence the electrons will describe circles of radius r given by

(5.28)
$$evB = \frac{mv^2}{r}$$

where m is the electronic mass. The angular velocity, ω, is given by

(5.29)
$$\omega = \frac{v}{r} = \frac{eB}{m}.$$

Hence the angular velocity is independent of the transverse velocity, v, and all the electrons will describe circles of various radii but in equal time. All these circles are tangential to each other on the axis of the hole in the anode. The magnetic field, **B**, can therefore be adjusted so that all the electrons complete one or more circles in their motion to the screen. This enables the electron stream to be focused. The electrostatic-deflection plates can be used to produce deflection of the beam in the plane of the screen.

The interaction of electric and motional forces plays a central role in vacuum tubes, particle accelerators, and nuclear-fusion devices. We shall see in the next section that additional forces can be obtained by varying the magnetic field in time.

5.4 Faraday's law

The expression for the force on moving charges given in eqn (5.12) was derived by deduction from Ampère's experimental results, which showed the equivalence of current loops and magnets, and by the additional idea inherent in eqn (5.8) that a current consists of a stream of charges and a current element is equivalent to the charge forming part of such a stream. This deduction involved Newton's law of action and reaction. Ampère was a dedicated follower of Newton and his results fit into the framework of Newtonian mechanics with its emphasis on action at a distance between material particles.

There are serious limitations in this approach which concern the effect of time. The argument was based on constant magnetic fields and constant currents. Also, the action between the particles and current elements was considered to be similar to the action in constant gravitational fields. Any changes would require infinite velocities of propagation, so that the action and reaction would be equal irrespective of the distance between the particles.

These difficulties were overcome by Michael Faraday, who put forward views which led to the overthrow of action at a distance theories and ultimately to the overthrow of Newtonian mechanics by Einstein's relativity theory. Faraday felt uneasy about the notion that physical systems could be divided into individual particles, whose effects could be added. His guiding principle was 'connectedness'. In our language, he felt that the primary quantity was the field with its tubes of flux and distributed energy and force. The field belonged to a system as a whole and not to isolated parts of a system. Moreover the field was a real substance, and its changes involved motion and time. Faraday, unlike Ampère, knew very little mathematics and relied entirely on experimentation. He had very sharp powers of observation and paid attention to transient effects, which had been disregarded by other experimenters. This led him in 1831 to the discovery of the transformer effect. He arranged two coils on an iron ring. The primary coil was connected to a voltaic cell and the secondary coil to a galvanometer. Currents were observed to flow in the secondary when the current in the primary was altered. The secondary currents were transient, and no current flowed in the secondary when the current in the primary was constant. Faraday attributed these results to the action of the magnetic flux in the iron ring, which linked the two coils. The idea of magnetic linkage was fundamental. Faraday had observed that, whereas electric flux terminated on conducting surfaces, magnetic flux was disposed in linkages around the current.

Since the secondary current requires an electric field to drive it, Faraday's law can be written as

(5.30)
$$\oint \boldsymbol{E} \cdot \mathrm{d}\boldsymbol{l} = -\frac{\mathrm{d}\Phi}{\mathrm{d}t}.$$

The negative sign embodies the experimental result that the secondary current was found to produce a flux opposing the change of flux in the ring.

This closed-loop integral of the electric field was met in eqns (5.15) and (5.18), where this voltage was called the electromotive force, or e.m.f., to distinguish it from the potential difference. It appears that there are two kinds of e.m.f.; one is caused by motion across a magnetic field, and the other is caused by a change in flux linkage. However, the motional effect is also related to flux linkage. Consider Fig. 5.2, where the rate of change of flux linkage is vBl, and where the sign of the e.m.f. and its resultant current is such as to oppose the increase of the flux linked. Hence the motional e.m.f. is a special case of eqn (5.30). Faraday carried out very extensive experiments and found that identical effects were obtained by relative movements of the primary and secondary coil, changes of currents, or changes of the shape of the coils. Equation (5.30) always agreed with the observed result (although Faraday did not use equations!).

Now consider a conducting loop surrounding a changing flux, as shown in Fig. 5.10. The flux is perpendicular to the plane of the loop and is directed upwards. An e.m.f. will be induced in the direction of the arrows. If the arrangement of the flux and loop is symmetrical, the electric field, E, of the e.m.f. will be constant around the loop. A current will flow in accordance with Ohm's law; that is,

(5.31)
$$\oint E \cdot dl = IR.$$

Now suppose that the loop is broken, so that no current can flow. The e.m.f. field will now act on the free charges in the loop, and they will be displaced. As a result, there will be a surface-charge distribution on the conducting loop, as indicated in Fig. 5.11. The surface charges will be associated with a conservative electric field and there will be a potential difference between the open ends of the conductor. Inside the conductor, this electric field will cancel the electric field of the e.m.f. If the gap between the open ends is

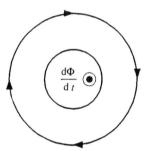

Fig. 5.10 A conducting loop surrounding a changing flux.

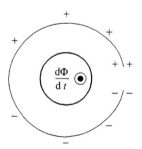

Fig. 5.11 Surface charges on a conducting loop.

small, the entire e.m.f. will be equal to the potential difference, which can be measured by a voltmeter. Suppose now that the ring is displaced from the symmetrical position of Fig. 5.10 to the position of Fig. 5.12. The closed e.m.f. integral is unchanged because the change of the flux linkage is the same. Hence the current in the closed ring is still given by eqn (5.31). However, the electric field associated with the e.m.f. is no longer constant around the ring. Nevertheless, Ohm's law requires that, locally, the electric field driving the current must be the same. This condition is met by the surface-charge distribution indicated in Fig. 5.13, which reinforces the e.m.f. at the right-hand side and reduces it at the left-hand side of the ring. The closed line integral of the conservative field of the surface charges is zero. Its action is to distribute the e.m.f. field uniformly.

Next, consider a conductor of several insulated loops surrounding the changing flux. Figure 5.14 shows two loops. This arrangement can be considered as two loops in series, that is, connected end to end. The

Fig. 5.12 Current flow in a displaced conductor.

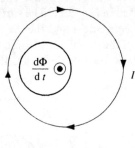

Fig. 5.13 Surface charges on a conductor.

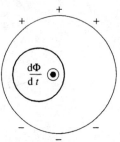

Fig. 5.14 Two loops surrounding a changing flux.

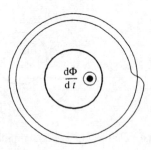

combined loop encircles the flux twice, so the e.m.f. will be twice its previous value. If there is a gap in the conductor, as in Fig. 5.11, the surface charges will double the potential difference. In general for N loops

(5.32)
$$\oint E \cdot dl = -N \frac{d\Phi}{dt}.$$

It is not necessary that N should be whole number. For example half a turn (symmetrically arranged) will give half the e.m.f. of a complete turn.

5.5 The transformer

The fact that the e.m.f. is nonconservative means that energy can be exchanged by its means. For example, the heat energy associated with Ohm's law can be supplied by the changing flux. Clearly, this flux cannot increase indefinitely, but it will have to alternate. But energy will still be supplied irrespective of the direction of the current. Hence energy will be supplied throughout a cyclic process.

This energy need not be used for heating, but it can be used for any kind of process requiring electrical energy. A very important device making use of this principle is the transformer. Consider two coils linked by a changing magnetic flux, Φ, as illustrated in Fig. 5.15. G is a generator supplying the current I_1 and L is a load which is fed with the current I_2. We have the following relationships

(5.33)
$$V_1 = \text{e.m.f.} = -N_1 \frac{d\Phi}{dt},$$

(5.34)
$$V_2 = \text{e.m.f.} = -N_2 \frac{d\Phi}{dt}.$$

Hence

(5.35)
$$\frac{V_1}{V_2} = \frac{N_1}{N_2}.$$

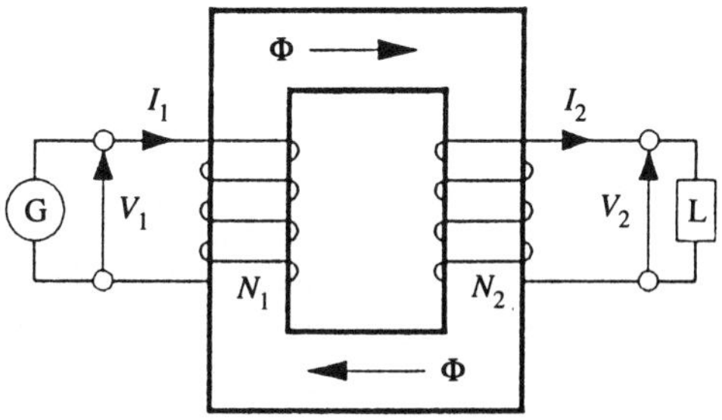

Fig. 5.15 A transformer.

The transformer changes the potential differences at the terminals of its coils in proportion to the turns ratio. Also

(5.36)
$$\oint \boldsymbol{H} \cdot \mathrm{d}\boldsymbol{l} = N_1 I_1 - N_2 I_2 .$$

In an 'ideal' transformer it can be assumed that the flux, Φ, needs a negligible field, H, to drive it. Then

(5.37)
$$N_1 I_1 - N_2 I_2 = 0$$

and

(5.38)
$$\frac{I_1}{I_2} = \frac{N_2}{N_1}$$

so that

(5.39)
$$V_1 I_1 = V_2 I_2$$

and the output is equal to the input power, as would be expected, because the resistances of the coils have been neglected. In an actual transformer, V_1 would have to be slightly larger than the e.m.f. in coil 1, and similarly V_2 would be slightly smaller than the e.m.f. in coil 2.

Some care has to be taken with the signs of the e.m.f.s since they depend on the direction of the flux relative to the direction in which the coils are wound. In Fig. 5.15 the coils are wound in opposite directions and the fluxes are also in the opposite direction.

A further complication in a real transformer arises from the phenomenon of *flux leakage*. This concept was discussed in Sections 3.6 and 3.7. In general, the flux linking the two coils will not be strictly equal. If, for example, the flux path is provided by an iron core, some of the flux will take the shorter path through the air as indicated in Fig. 5.16. However, the leakage flux is generally only a few per cent of the mutual flux.

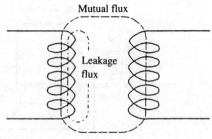

Fig. 5.16 Mutual flux and leakage flux.

5.6 Self-inductance and mutual inductance

In the discussion of the motional e.m.f. in Section 5.2 and of the transformer e.m.f. in Section 5.5 it was assumed that the magnetic flux was provided by an independent current system. This enabled the e.m.f. induced in a coil to be separated from the current in that coil. Now consider the interaction of the flux and the current in the same coil.

Let the coil have a resistance R and let there be a flux linkage, Φ, when a current I flows. It will be remembered that I produces a magnetic field H and that H is associated with B. For constant μ, B is proportional to H. The flux linkage is proportional to B, so that Φ is proportional to I. The relation of Φ to I depends on the geometry of the coil. The self-inductance of the coil, L, is defined by

(5.40)
$$L = \frac{\Phi}{I}.$$

Let the coil be connected to a generator, which applies a constant potential difference, V, between the ends of the coil. This potential difference will have to overcome the e.m.f. caused by the changing flux linkage as the current increases. It will be remembered that the e.m.f. will act so as to oppose the change in the flux, and V will also have to provide the ohmic drop caused by the resistance. Thus

(5.41)
$$V = \frac{d\Phi}{dt} + RI.$$

Hence

(5.42)
$$V = L\frac{dI}{dt} + RI.$$

The energy supplied by the generator is

(5.43)
$$U = \int_0^t VI\,dt = \tfrac{1}{2}LI^2 + \int_0^t RI^2\,dt.$$

The second term gives the energy converted to heat. The first term is the energy associated with the flux linkage. It has already been noted that this magnetic energy is analogous to kinetic energy and that it represents the distributed kinetic energy of the current. A comparison of eqn (5.42) with mechanical motion is interesting. For an equivalent mechanical system we can write

(5.44)
$$F = m\frac{dv}{dt} + kv.$$

The second term represents viscous friction and the first term the inertial force of a mass. The kinetic energy $\frac{1}{2}mv^2$ can be compared with $\frac{1}{2}LI^2$, and L is analogous to inertial mass. The chief difference is that the kinetic energy of mass is inherent in the mass whereas the kinetic energy of the current is distributed in space.

The relationship between the current and time is given in Fig. 5.17. Algebraically this is given by

$$(5.45) \qquad I = \frac{V}{R}\left(1 - e^{-Rt/L}\right).$$

The steady-state current is defined by Ohm's law, but the steady state cannot be reached instantaneously. The acceleration of the charges, that is, the slope of the graph of the current, is given at zero time by

$$(5.46) \qquad \left(\frac{dI}{dt}\right)_{t=0} = \frac{V}{L}.$$

In the mechanical analogy of a train being pulled by an engine of constant tractive force, the acceleration from a standstill is determined by the inertial mass of the train and its final speed by frictional losses. The duration of the transient is governed by the index of the exponential, which can be written $e^{-t/\tau}$, where τ is the time constant given by

$$(5.47) \qquad \tau = \frac{L}{R}.$$

If the inertia is dominant, τ is large; and if the friction is dominant, τ is small.

All these results show that the magnetic field energy interacts with the electric field driving the current in a similar way to the interaction of a moving mass and an applied force as given by Newton's laws of motion. It shows that time is an essential factor in the process and that instantaneous action at a distance between stationary particles does not provide an adequate description.

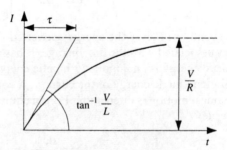

Fig. 5.17 Build-up of current in the circuit.

Now consider the case of two coils having currents I_1 and I_2, respectively. The flux linkage with coil 1 will in general depend on both currents. The mutual linkage can be defined in terms of a mutual inductance, M, given by

$$(5.48) \qquad M = \frac{\Phi_1}{I_2},$$

where Φ_1 is the flux linkage with coil 1 due to I_2. Equation (5.41) applies as before to coil 1, but eqn (5.42) has to be modified because the flux linkage is

$$(5.49) \qquad \Phi = LI_1 + MI_2.$$

The energy supplied by the generator is now

$$(5.50) \qquad U = \tfrac{1}{2}LI_1^2 + MI_1I_2 + \int_0^t RI_1^2 \, dt.$$

The second term is the mutual kinetic energy of the two currents I_1 and I_2. Since it is symmetrical in I_1 and I_2, the mutual inductance can also be written

$$(5.51) \qquad M = \frac{\Phi_2}{I_1},$$

where Φ_2 is the flux linkage with the current system 2 due to the current system 1. Sometimes it is convenient to write the mutual inductance as M_{12} or M_{21}. We have shown that

$$(5.52) \qquad M_{12} = M_{21}.$$

Another notation is to write $L = L_{11}$ and $M = L_{12}$. It is straightforward to extend the treatment to an arbitrary number of coils. The total magnetic energy of a system of coils is given by

$$(5.53) \qquad U = \sum_m \tfrac{1}{2}L_{mm}I_m^2 + \sum_{\substack{m,n \\ m \neq n}} \tfrac{1}{2}M_{mn}I_mI_n.$$

5.7 The skin effect and eddy currents

It was shown in the last section that magnetic-field energy has an inertial nature, so that time is required to establish a steady current. From Ohm's law, steady currents in conductors of uniform cross section are distributed uniformly at a constant current density. However, in most practical cases currents are alternating at various frequencies so that the magnetic energy is changing all the time, and the current distribution will depend not only on the resistance but also on the inductance. We shall expect to find that the energy supplied to a conductor will penetrate only a limited distance into the conducting material because each positive half-cycle is followed by a negative half-cycle. The current density will be a maximum near the surface of a conductor and if the conductor is thick enough there will be negligible current in its central region. This phenomenon is known as the *skin effect*, and we shall now investigate it carefully.

Because the current is distributed nonuniformly we must deal with its local distribution. That means that a field description is needed.

The electric field energy is negligibly small, so there are two kinds of energy involved: the ohmic loss and the magnetic energy. These are associated with the conductivity, σ, and the permeability, μ, respectively. We have the constitutive equations

(5.54)
$$\sigma E = J$$

and

(5.55)
$$\mu H = B.$$

Also, there is the relationship between the current and the magnetic field

(5.56)
$$\oint H \cdot dl = I = \iint J \cdot dS$$

and the relationship between the changing magnetic flux and the electric field

(5.57)
$$\oint E \cdot dl = -\frac{d\Phi}{dt} = -\frac{d}{dt}\iint B \cdot dS.$$

In order to use eqns (5.56) and (5.57) locally we need small loops surrounding small areas. In vector notation

(5.58)
$$\operatorname{curl} H = J$$

(5.59)
$$\operatorname{curl} E = -\frac{\partial B}{\partial t}.$$

To simplify the initial treatment, consider an immensely thick conductor, thick enough for the energy density to be negligible deep in the interior. With such a thick conductor the outside surface can be treated as locally flat. Also it can be assumed that the current flows in one direction only, parallel to the surface, as shown in Fig. 5.18. From eqns (5.58) and (5.59)

(5.60)
$$-\frac{\partial H_y}{\partial z} = J_x$$

and

(5.61)
$$+\frac{\partial E_x}{\partial z} = -\frac{\partial B_y}{\partial t}.$$

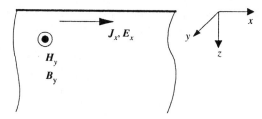

Fig. 5.18 Current flow in a semi-infinite conductor.

Hence from eqn (5.60)

(5.62)
$$-\frac{1}{\mu}\frac{\partial B_y}{\partial z} = \sigma E_x$$

and

(5.63)
$$-\frac{\partial^2 B_y}{\partial z\,\partial t} = \mu\sigma\frac{\partial E_x}{\partial t}.$$

From eqn (5.61)

(5.64)
$$-\frac{\partial^2 B_y}{\partial t\,\partial z} = \frac{\partial^2 E_x}{\partial z^2}$$

so that

(5.65)
$$\frac{\partial^2 E_x}{\partial z^2} = \mu\sigma\frac{\partial E_x}{\partial t}.$$

This equation relates the distance z into the conductor to the time t. It is called the diffusion equation because it describes the diffusion of energy into the conductor.

Consider harmonic time variation of the type sin ωt or cos ωt, where ω is the angular frequency. The mathematics can be simplified by using $e^{j\omega t}$. Hence $E = E(z)\,e^{j\omega t}$ and

(5.66)
$$\frac{\partial^2 E(z)}{\partial z^2} = j\mu\sigma\omega\,E(z)$$

where $e^{j\omega t}$ has been cancelled from both sides of the equation. Now

(5.67)
$$\frac{\partial^2 E}{\partial z^2} = a^2 E$$

has the solution

(5.68)
$$E = Ae^{az} + Be^{-az}$$

where A and B are constants. In our example $A=0$, because the field decreases with the distance z. Hence

$$(5.69) \qquad E(z)=B \exp[-(j\mu\sigma\omega)^{1/2}z].$$

Now

$$(5.70) \qquad \sqrt{j}=\frac{1+j}{\sqrt{2}}$$

and

$$(5.71) \qquad B=E_s e^{j\omega t}$$

where E_s is the magnitude of the surface electric field. Hence

$$E=E(z)\exp(j\omega t)$$

$$(5.72) \qquad = E_s \exp[-(\tfrac{1}{2}\mu\sigma\omega)^{1/2}z]\exp\{j[\omega t-(\tfrac{1}{2}\mu\sigma\omega)^{1/2}z]\}.$$

For simplicity put

$$(5.73) \qquad \Delta=\left(\frac{2}{\mu\sigma\omega}\right)^{1/2}$$

so that

$$(5.74) \qquad E=E_s e^{-z/\Delta}e^{j(\omega t-z/\Delta)}$$

and finally

$$(5.75) \qquad E=E_s e^{-z/\Delta} \cos(\omega t-z/\Delta)$$

where a phase has been chosen which corresponds to $\cos \omega t$ at the surface.

Equation (5.75) deserves close attention. The electric field, E, decays exponentially in magnitude with distance into the conductor. This decay depends on Δ, which is a length (since z/Δ must be a number in order to be dimensionless). The decay is sharper the larger the ratio z/Δ becomes; that is, the smaller Δ becomes. A steep decay therefore depends on large values of the permeability, the conductivity, and the frequency. Not only does the field decay exponentially, but it also changes phase, so that at a certain depth the field may be in the opposite direction to the field at the surface. This fits in with the fact that the energy at that depth is due to the previous half-cycle at the surface.

We have written the field in terms of E, but it is easy to show that we could equally well have used J, H or B

$$(5.76) \qquad J=J_s e^{-z/\Delta} \cos(\omega t-z/\Delta),$$

$$(5.77) \qquad H=H_s e^{-z/\Delta} \cos(\omega t-z/\Delta),$$

$$(5.78) \qquad B=B_s e^{-z/\Delta} \cos(\omega t-z/\Delta).$$

Of course J and E will be in phase with each other, and so will B and H. But J and H, and E and B will have components which are in phase quadrature.

Let us now insert some numbers to determine Δ in some practical situations. For copper at 50 Hz, assuming a conductivity of $\sigma = 0.5 \times 10^8$ S m^{-1}

$$(5.79) \qquad \Delta = \left(\frac{2}{4\pi \times 10^{-7} \times 0.5 \times 10^8 \times 100\pi} \right)^{1/2} = 0.01 \text{ m} = 10 \text{ mm}.$$

At a depth $z = \Delta$, the fields are reduced to 36.8 per cent of the surface value; and at a depth of $z = 4\Delta$, they are reduced to 1.8 per cent. A conductor of thickness 8Δ could be regarded as 'infinitely thick'.

For copper at 5 kHz, $\Delta = 1$ mm, and at 50 MHz, $\Delta = 0.01$ mm. Clearly all conductors at radio frequencies are infinitely thick.

For iron at 50 Hz, $\mu_r = 3000$ and $\sigma = 0.106 \times 10^8$ S m^1 so $\Delta = 0.4$ mm. Hence, iron is 'thick' even at power frequencies, unless it is used in thin sheets.

The entire phenomenon is governed by the length Δ, which is called the *skin depth* or *penetration depth*. Our treatment shows that the field (and the energy) penetrates into conductors well below that depth, but most of the energy is contained in the surface region.

Now consider the ohmic loss in a thick conductor. The power per unit surface area is given by

$$(5.80) \qquad P = \int_0^\infty \frac{|J|^2}{2\sigma} \, dz.$$

The upper limit of the integral has been chosen as infinity in order to simplify the mathematics. The symbol $|J|$ means that we are using the magnitude of J, because the phase makes no difference to the local power loss.

$$(5.81) \qquad |J| = J_s e^{-z/\Delta}.$$

Hence

$$(5.82) \qquad P = \frac{J_s^2 \Delta}{4\sigma}.$$

The complex amplitude of the total current per unit distance in the y-direction is given by

$$(5.83) \qquad I = \int_0^\infty J_s e^{-(1+j)z/\Delta} \, dz = \frac{J_s \Delta}{1+j}.$$

Hence the amplitude of the alternating current is

$$(5.84) \qquad |I| = \frac{J_s \Delta}{\sqrt{2}}.$$

Hence

(5.85)
$$P = \frac{|I|^2}{2\sigma\Delta},$$

and the equivalent resistance is given by

(5.86)
$$R = \frac{1}{\sigma\Delta}.$$

This means that the ohmic loss of the actual current is equal to the loss that would be associated with a current of equal magnitude uniformly distributed in a conductor of thickness Δ. Equation (5.83) shows that the phase of the total current lags by $\pi/4$ behind the phase of the surface current density. Since the surface current is in phase with the surface electric field, this means that the total current lags $\pi/4$ behind the applied voltage. Hence the inductive reactance is equal to the resistance and the magnetic energy is equal to the energy loss. The power factor is $\cos \pi/4 = 0.7071$.

The example of an infinitely thick conductor is somewhat artificial; but if the thickness is of the order of 10Δ the assumption is fully justified, because the field at the centre would be of the order of $e^{-5} = 0.0067$ (0.67 per cent) and the energy density would be of the order $e^{-10} = 0.000\ 045$ (0.0045 per cent).

However, not all conductors are thick compared to the skin depth. It is wasteful to use conductors which carry no current or only very little current over part of their cross-section. A tubular conductor would be more economical in its use of material but would be more expensive to produce. More generally, conductors are subdivided into strands as segments which are connected in parallel. This is particularly useful at power frequencies.

Let us analyse the behaviour of thin conductors in which the thickness is less than the skin depth. This implies that the energy will penetrate the entire conductor material and there will be relatively little variation in magnitude or phase. The variation of the current can therefore be treated as a perturbation of the current distribution for steady currents. Notice that as $\omega \to 0$, $\Delta \to \infty$ since $\Delta = (2/\mu\sigma\omega)^{1/2}$, so that the steady current distribution for $\omega = 0$ will give the ratio, thickness/skin-depth $= 0$. But the skin effect can be expected to be quite small as long as the skin depth is several times larger than the thickness.

As an example, consider a conductor in the form of a flat sheet of thickness $2b$, as illustrated in Fig. 5.19. Let us take x, y, z coordinates with their origin on the central plane. Let J_0 be the uniform current density in the x-direction. This will give rise to a magnetic field H_0 given by eqn (5.58) as

(5.87)
$$-\frac{\partial H_0}{\partial z} = J_0.$$

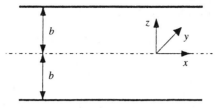

Fig. 5.19 Current flow in a flat sheet.

Hence

(5.88)
$$H_0 = -J_0 z.$$

There is no constant of integration because H_0 is zero for $z=0$. Hence

(5.89)
$$B_0 = -\mu J_0 z$$

The variation, in time, of B_0 with the angular frequency, ω, induces an additional electric field, E_1, given by

(5.90)
$$\frac{\partial E_1}{\partial z} = -j\omega B_0,$$

and E_1 drives an additional current of density J_1, whence

(5.91)
$$\frac{\partial J_1}{\partial z} = -j\mu\sigma\omega H_0 = j\mu\sigma\omega J_0 z.$$

Hence

(5.92)
$$J_1 = \tfrac{1}{2}j\mu\sigma\omega J_0 z^2 + K.$$

Assume that the actual total current per metre in the y-direction is given by

(5.93)
$$I = 2J_0 b$$

Hence the integral of J_1 across the conductor must be zero, J_1 is an *eddy current*.

(5.94)
$$0 = \int_{-b}^{+b} J_1 \, dz = \tfrac{1}{3}j\mu\sigma\omega b^3 J_0 + 2bK.$$

Hence

(5.95)
$$K = -j\frac{b^2}{3\Delta^2} J_0$$

and

(5.96)
$$J_1 = \frac{jJ_0}{\Delta^2}\left(z^2 - \frac{b^2}{3}\right).$$

Now consider the series

(5.97)
$$J = J_0 + J_1 + J_2 + \ldots J_k + \ldots$$

and

(5.98)
$$B = B_0 + B_1 + B_2 + \ldots B_k + \ldots,$$

J_0 gives rise to B_0, B_0 to J_1, J_1 to B_1, etc. Notice that J_1 is of the order $J_0 b^2/\Delta^2$; and it is found that J_2 is of the order $J_1 b^2/\Delta^2$. Since $\Delta > b$, this series converges rapidly.

Now consider the ohmic power loss given by eqn (5.80). Consider only J_0 and J_1

(5.99)
$$J^2 = J_0^2 + 2J_0 J_1 + J_1^2,$$

J_0 and J_1 are in phase quadrature and will give no mutual loss. Hence

(5.100)
$$P = \int_{-b}^{+b} \frac{|J_0|^2 + |J_1|^2}{2\sigma} \, dz = \frac{J_0^2 b}{\sigma} + \frac{4}{45} J_0^2 b \left(\frac{b}{\Delta}\right)^4.$$

The equivalent a.c. resistance is given by

(5.101)
$$R_{\text{a.c.}} = R_{\text{d.c.}} \left(1 + \frac{4}{45} \frac{b^4}{\Delta^4}\right).$$

The next approximation involving J_2 will give a term in b^8/Δ^8, which can generally be neglected.

For a given conductivity and permeability, $(b/\Delta)^4$ varies as ω^2. Hence

(5.102)
$$R_{\text{a.c.}} = R_{\text{d.c.}}(1 + k\omega^2).$$

This relationship applies to conductors of any shape, if only the first-order eddy current, J_1 is considered. It should be compared with

(5.103)
$$R_{\text{a.c.}} = \frac{1}{\sigma\Delta} = k\omega^{1/2}$$

for thick conductors. The two relationships are illustrated in Fig. 5.20. The solution for thin conductors is called the *resistance-limited* condition, and

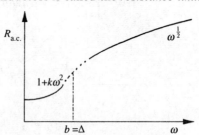

Fig. 5.20 Variation of the a.c. resistance with the frequency.

the solution for thick conductors is called the *inductance-limited* condition. The latter term is somewhat misleading, because in thick conductors the inductive reactance is equal to the resistance.

The two solutions are also described as '*low frequency*' and '*high frequency*', but it must be remembered that the criterion used is not the frequency by itself but the ratio of the thickness to the skin depth. It is best to distinguish between thick and thin conductors. The dotted part of the curve in Fig. 5.20 is the region where neither approximation is correct. The exact shape of the conductor then becomes important. However, it is found that this intermediate behaviour occurs in a very limited region of thickness. The dotted curve can often be formed by interpolation between the approximate thick and thin behaviour.

5.8 Electrokinetic momentum and the vector potential

Faraday's law gives the e.m.f. in terms of the change of flux linkage. The e.m.f. is the closed line-integral of the electric field strength, E, and the question arises whether E itself, as well as its integral, could be determined. We have already seen from eqn (5.13) in Section 5.2 that the motional component can be found from

$$E = v \times B.$$

Let us now try to find the electric field due to the transformer action, where E is to be measured by a stationary probe. Now

(5.104)
$$\oint E \cdot dl = -\frac{\partial \Phi}{\partial t} = -\iint \frac{\partial B}{\partial t} \cdot dS.$$

We want to convert the surface integral on the right-hand side into a line integral. Suppose we write

(5.105)
$$\Phi = \oint A \cdot dl.$$

Then

(5.106)
$$\oint E \cdot dl = -\oint \frac{\partial A}{\partial t} \cdot dl$$

and

(5.107)
$$E = -\frac{\partial A}{\partial t} - \text{grad } \phi.$$

The gradient field has to be introduced because the line integral of a gradient field is zero. Physically there might be an electrostatic field present, but grad ϕ is perfectly general and its introduction shows that the vector A

has not been uniquely specified by eqn (5.105). We shall return to this problem in the next chapter, but meanwhile consider the term

$$(5.108) \qquad E = -\frac{\partial A}{\partial t}.$$

The field E is the force on a unit charge. In mechanics force is equal to the rate of change of momentum. This suggests an anology with A. We already regard the magnetic field as describing the kinetic energy associated with electric currents. Maxwell, therefore, called A the electrokinetic momentum. There is of course a difference between this kind of momentum and mechanical momentum. In mechanics the force and the momentum are related to a local mass; but, in electromagnetism, A is the combined mutual momentum of the entire system referred to the point at which E is measured.

In terms of a small loop, eqn (5.105) can be written

$$(5.109) \qquad \text{curl } A = B.$$

Since

$$(5.110) \qquad B = \mu H$$

and

$$(5.111) \qquad \text{curl } H = J,$$

A can be written in terms of J as

$$(5.112) \qquad \text{curl curl } A = \text{curl } \mu H = \mu J.$$

The differential operator curl curl can be written in x, y, z coordinates and the terms can be rearranged to show that

$$(5.113) \qquad \text{curl curl } A = \text{grad div } A - \nabla^2 A,$$

where $\nabla^2 A$ is defined by

$$(5.114) \qquad \nabla^2 A = \hat{x}\nabla^2 A_x + \hat{y}\nabla^2 A_y + \hat{z}\nabla^2 A_z.$$

We can put

$$(5.115) \qquad \text{div } A = 0$$

because a divergence field is conservative, and eqn (5.109) defines only the nonconservative part of A. Hence, finally,

$$(5.116) \qquad \nabla^2 A = -\mu J$$

or

$$(5.117) \qquad \left. \begin{aligned} \nabla^2 A_x &= -\mu J_x, \\ \nabla^2 A_y &= -\mu J_y, \\ \nabla^2 A_z &= -\mu J_z. \end{aligned} \right\}$$

These equations can be compared with Poisson's equation in electrostatics, where

(5.118)

$$\nabla^2 V = -\frac{\rho}{\varepsilon}.$$

This comparison shows that A behaves similarly to V. The scalar potential, V, is related to the charge distribution, ρ, and the vector A is related to the current distribution, J. Hence A is generally called the vector potential. The name is slightly unfortunate because potential usually means potential energy, and energy is always a scalar and it cannot have direction. Hence a vector potential energy is impossible. Nevertheless, the term vector potential is well established, and Maxwell's term, electrokinetic momentum, is seldom used in spite of the useful analogy with the momentum of moving masses.

Exercises

5.1 In a certain cathod-ray tube, focusing is achieved by means of a uniform axial magnetic field produced by a long solenoid. The flux density, B (in T), is related to the current I (in A), by the relation $B = 5 \times 10^{-3} I$. It is found that a sharply defined spot is obtained for successive values of the current of 0.9, 1.7, 2.6, and 3.5 A. Estimate the transit time between the anode hole and the screen. (*Answer* 8.16×10^{-9} s.)

5.2 Figure 5.21 shows an axially magnetized cylindrical magnet which can rotate on its axis. A conducting disc is mounted coaxially with the magnet and the disc can rotate independently of the magnet. A galvanometer is connected to the centre of the disc and to its circumference by means of a sliding contact. Apply Faraday's law to decide whether the galvanometer will deflect when: (a) the disc is

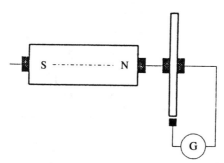

Fig. 5.21 A homopolar generator.

rotated and the magnet is stationary, (b) the disc is stationary and the magnet rotates, and (c) both the disc and the magnet rotate in the same direction and at the same angular velocity.
(*Answer* (a) yes, (b) no, and (c) yes.)

5.3 In the circuit shown in Fig. 5.22, let the end wires each have a total resistance $R = 25\ \Omega$, with no resistance in the horizontal wires. The uniform magnetic field B has a magnitude of 0.3 T and it varies sinusoidally in time with a frequency of 400 Hz. Find the readings on the voltmeter and the ammeter.
(*Answer* 6.66 $V_{r.m.s.}$ and 0.267 $A_{r.m.s.}$.)

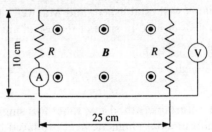

Fig. 5.22 The circuit for Exercise 5.3.

5.4 Figure 5.23 shows a sliding bar moving at a constant velocity of 4 m s^{-1} through a uniform field $B = 0.5$ T. Let $R = 5\ \Omega$. (a) Find V_{12} and V_{34}. (b) Find I_a and I_b. (c) What force is required to move the bar? (d) Show that the work done in moving the bar from one end to the other is equal to the energy delivered to R.
(*Answer* (a) -0.2, -0.2 V, (b) 0, -0.04 A, (c) 0.002 N, and (d) 0.001 J.)

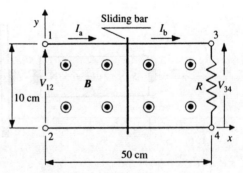

Fig. 5.23 The circuit for Exercises 5.4, 5.5, and 5.6.

5.5 The sliding bar in Fig. 5.23 oscillates back and forth between positions 1–2 and 3–4 (that is, between $x=0$ and $x=50$ cm) in a sinusoidal manner 30 times a second. Assume it is at the left end at $t=0$, and determine I_b as a function of time if $R=0.2\ \Omega$ and $B=1.2$ T.
(*Answer* $-28.3\sin 60\pi t$ A.)

5.6 Find $I_b(t)$ for the circuit of Fig. 5.23 if the location of the sliding bar is specified by $x=0.25(1-\cos\omega t)$ m, while $B=0.4\cos\omega t$ T and $R=0.1\ \Omega$.
(*Answer* $0.1\ \omega(\sin 2\omega t-\sin\omega t)$.)

5.7 The limb of a transformer has a circular cross section of diameter d. It carries an alternating magnetic flux of peak amplitude Φ, of frequency f and is of uniform flux density across the section. Estimate the electric field due to the magnetic flux (magnitude, phase, and direction) if the other parts of the transformer are sufficiently distant for their effects to be neglected.
(*Answer* $E_\Phi=-\Phi f/d$.)

5.8 A thin conducting ring of mean diameter $2d$ and cross section s surrounds the transformer limb of Exercise 5.7 and a current I flows in the ring. Obtain an expression for this current if the conductivity of the ring is σ. Explain why there will be no potential difference between parts of the ring if it is concentric with the transformer limb, but that a potential difference will appear when the centre of the ring is displaced from the axis of the transformer limb.
(*Answer* $I=-\Phi\sigma sf/d$.)

5.9 A circular copper disc is placed centrally inside a solenoid and spun about its axis which is coincident with the axis of the solenoid. The solenoid has a length of 30 cm and a mean diameter of 12 cm. It is uniformly wound with 4000 turns. The diameter of the disc is 8 cm. Calculate the e.m.f. induced between the centre and rim of the disc when it is rotated at 3000 rev min^{-1} and the current in the solenoid is 5 A.
(*Answer* 19.5 mV.)

5.10 A long solenoid with inductance L in air has a radius of cross section a. A single-turn loop of radius r is centred coaxially with the solenoid. Find the mutual inductance in terms of L if: (a) $r=\frac{1}{2}a$, (b) $r=a$, and (c) $r=2a$.
(*Answer* (a) $L/4$, (b) L, and (c) L.)

5.11 Show that the internal inductance per unit length of a nonmagnetic cylindrical wire of radius a carrying a uniformly distributed current I is $\mu_0/8\pi$ H m^{-1}.

5.12 (a) Find the mutual inductance in air between a long straight filament and a coplanar square loop, 10 cm on each side, with its nearest edge parallel to and 10 cm from the filament. (b) Find the mutual inductance if the square loop is rotated 45° in its plane about its centre. (*Answer* (a) 13.86 nH, and (b) 13.88 nH.)

5.13 Define the term *skin depth* applied to the flow of alternating current in a conductor. Discuss what is meant by *thin* and *thick* conductors.

5.14 Show that the a.c. resistance of a thin conductor of circular cross section is given by

$$R_{\text{a.c.}} = R_{\text{d.c.}}\left(1 + \frac{1}{48}\frac{a^4}{\Delta^4}\right),$$

where a is the wire radius and Δ is the skin depth. Show that the a.c. resistance of copper wire of 1 cm diameter at 100 Hz is approximately half a per cent higher than the d.c. resistance. What is the increase at 400 Hz?
(*Answer* 8 per cent.)

5.15 Discuss the terms *inductance-limited* and *resistance-limited* eddy currents. Show that under certain conditions the a.c. resistance of a circular conductor of radius a is given by the expression $R_{\text{a.c.}} = R_{\text{d.c.}}a/2\Delta$. What are these conditions? In a conductor to which these conditions apply it is proposed to reduce the loss by cooling. If the resistivity is reduced by 99 per cent by how much will the losses be reduced?
(*Answer* by 90 per cent.)

5.16 Alternating current flows parallel to the surface of a semi-infinite slab of conducting material. At what depth is the current density reduced to (a) 10 per cent and (b) 1 per cent of the surface value? At what depth is the phase of the current density (c) 90°, and (d) 180° lagging on the surface value?
(*Answer* (a) 2.3Δ, (b) 4.6Δ, (c) 1.57Δ, and (d) 3.14Δ.)

Electromagnetic radiation

6

6.1 Displacement current and Maxwell's equations

Faraday's law and the analogy between magnetic and kinetic energy have introduced us to the idea that time is involved in electromagnetic processes. The relationship between the magnetic field strength and the current given by

$$\oint H \cdot dl = I$$

(6.1)

involves time in the motion of charges, but it does not allow for changes of current. It will be remembered that this equation was derived from Ampère's equivalence theorem between constant currents and permanent magnets. As it stands, eqn (6.1) implies that the magnetic field, H, would change instantaneously with a change of current and this conflicts with Faraday's law. Moreover, the local form of eqn (6.1) gives

$$\operatorname{curl} H = J,$$

(6.2)

and since

$$\operatorname{div} \operatorname{curl} H \equiv 0$$

(6.3)

this implies that

$$\operatorname{div} J = 0,$$

(6.4)

so that eqn (6.1) implies that the current flow is continuous and unchanging. Clearly, eqn (6.1) applies only to direct currents, and if the current is alternating it can be only approximately correct.

There is another difficulty which relates to this equation. If current is a flow of charge, then all movements of charge should give a magnetic field. Equation (6.1) contains the conduction current, but omits the polarization current caused by the motion of charge in dielectrics. It is likely that we should add the polarization current and this would give

$$\operatorname{curl} H = J + \frac{\partial P}{\partial t}.$$

(6.5)

However, even this is incorrect. This problem was solved by James Clerk Maxwell, who argued that since electrical energy was not confined to dielectrics but could exist in free space the change of electric fields always had the nature of a polarization current. He proposed the correct equation, which is

(6.6)
$$\operatorname{curl} \boldsymbol{H} = \boldsymbol{J} + \frac{\partial \boldsymbol{P}}{\partial t} + \varepsilon_0 \frac{\partial \boldsymbol{E}}{\partial t},$$

hence

(6.7)
$$\operatorname{curl} \boldsymbol{H} = \boldsymbol{J} + \frac{\partial \boldsymbol{D}}{\partial t}.$$

He called $\partial \boldsymbol{D}/\partial t$ the *displacement current density*, since it is due to the displacement of doublets of charge in dielectrics and the displacement of electric energy in free space. If there is no polarization, the displacement is only that of the free-space electrical energy.

If eqn (6.3) is now applied, then

(6.8)
$$\operatorname{div} \boldsymbol{J} + \frac{\partial}{\partial t} \operatorname{div} \boldsymbol{D} = 0.$$

But

(6.9)
$$\operatorname{div} \boldsymbol{D} = \rho,$$

where ρ is the charge density, so that

(6.10)
$$\operatorname{div} \boldsymbol{J} = -\frac{\partial \rho}{\partial t}.$$

This is the continuity equation for electric charge. A reduction of charge with time gives rise to an outflow of current. Equation (6.4) is a special case of eqn (6.10), when there is no charge variation because the flow of current is constant. Similarly, eqn (6.2) is a special case of eqn (6.7).

Now consider the magnitude of the displacement current. If the time variation is harmonic, then

$$\left| \frac{\partial \boldsymbol{D}}{\partial t} \right| = \omega \varepsilon E.$$

Hence the ratio $|\boldsymbol{J}|$ to $|\partial \boldsymbol{D}/\partial t|$ is $\sigma/\omega\varepsilon$. In a conductor, $\varepsilon = \varepsilon_0$, which is of the order of 10^{-11}, so the ratio for a good conductor is of the order of $10^{18}/\omega$. This is enormous, and it can safely be assumed that the displacement current in a conducting material is negligible in comparison with the conduction current. In terms of energy, this means that the electric field energy is negligible compared with the magnetic field energy and the resistive loss. This is the assumption made in our treatment of the skin effect and eddy currents in the last chapter.

However, where there is no conduction current, the displacement current cannot be neglected. We then have

(6.11)
$$\text{curl } \boldsymbol{H} = \frac{\partial \boldsymbol{D}}{\partial t}.$$

This means that the magnetic field in a time-varying system has vortex sources in free space. Close to conduction currents, these sources will not greatly alter the magnetic field itself or the associated energy, but we shall show that at a distance these will become the dominant sources. The meanings of a *near field* and a *distant field* are also clarified later in this chapter. Meanwhile, notice that inductance calculations which are based on flux linkage with conduction current are near-field calculations because they neglect the displacement current.

Maxwell was the first investigator to use the various partial differential equations as a description of electromagnetic phenomena. Nowadays the term *Maxwell's equations* is synonymous with electromagnetism. We now list these equations:

(6.12)
$$\begin{cases} \text{curl } \boldsymbol{H} = \boldsymbol{J} + \dfrac{\partial \boldsymbol{D}}{\partial t}, \\[2mm] \text{curl } \boldsymbol{E} = -\dfrac{\partial \boldsymbol{B}}{\partial t}, \\[2mm] \text{div } \boldsymbol{D} = \rho, \\[2mm] \text{div } \boldsymbol{B} = 0, \\[2mm] \boldsymbol{D} = \varepsilon \boldsymbol{E}, \\[2mm] \boldsymbol{B} = \mu \boldsymbol{H}. \end{cases}$$

The current density, $\boldsymbol{J}$, and the charge density, ρ, are the sources of the field and they are connected by the continuity equation, eqn (6.10). The last two equations are called the *constitutive equations*. They give the basis of the method of tubes and slices, and they associate energy with the space in which the field acts.

6.2 Electromagnetic waves

There is a remarkable similarity between eqn (6.11) and the local form of Faraday's law, which is

(6.13)
$$\text{curl } \boldsymbol{E} = -\frac{\partial \boldsymbol{B}}{\partial t}.$$

Changing electric fields are vortex sources of magnetic fields and vice versa.

We can combine these two equations and eliminate either the electric or the magnetic field.

(6.14) $$\text{curl curl } \boldsymbol{H} = \text{curl } \frac{\partial \boldsymbol{D}}{\partial t} = \varepsilon \frac{\partial}{\partial t} \text{ curl } \boldsymbol{E} = -\varepsilon\mu \frac{\partial^2 \boldsymbol{H}}{\partial t^2}.$$

This can be simplified by using the vector identity

(6.15) $$\text{curl curl } \boldsymbol{H} = \text{grad div } \boldsymbol{H} - \nabla^2 \boldsymbol{H}.$$

In free space, the permeability is constant and equal to μ_0. Since div $\boldsymbol{B} = 0$ because there are no free poles, div $\mu_0\boldsymbol{H} = 0$; and therefore div $\boldsymbol{H} = 0$. Hence, from eqns (6.14) and (6.15)

(6.16) $$\nabla^2 \boldsymbol{H} = \mu\varepsilon \frac{\partial^2 \boldsymbol{H}}{\partial t^2}.$$

Also, in free space div $\boldsymbol{D} = 0$, and hence div $\boldsymbol{E} = 0$, so that

(6.17) $$\nabla^2 \boldsymbol{E} = \mu\varepsilon \frac{\partial^2 \boldsymbol{E}}{\partial t^2}.$$

Suppose $\boldsymbol{E}$ is in the x-direction, then eqn (6.17) becomes

(6.18) $$\nabla^2 E_x = \mu\varepsilon \frac{\partial^2 E_x}{\partial t^2}.$$

Since

$$\nabla^2 \equiv \frac{\partial^2}{\partial x^2} + \frac{\partial^2}{\partial y^2} + \frac{\partial^2}{\partial z^2}$$

then

(6.19) $$\frac{\partial^2 E_x}{\partial x^2} + \frac{\partial^2 E_x}{\partial y^2} + \frac{\partial^2 E_x}{\partial z^2} = \mu\varepsilon \frac{\partial^2 E_x}{\partial t^2}.$$

The divergence of E_x is $\partial E_x / \partial x$, and it is zero. Hence

(6.20) $$\frac{\partial^2 E_x}{\partial y^2} + \frac{\partial^2 E_x}{\partial z^2} = \mu\varepsilon \frac{\partial^2 E_x}{\partial t^2}.$$

To simplify the treatment consider an electric field which does not alter in the y-direction, so that E_x is constant in the xy-plane. Then

(6.21) $$\frac{\partial^2 E_x}{\partial z^2} = \mu\varepsilon \frac{\partial^2 E_x}{\partial t^2}.$$

From eqn (6.13),

(6.22) $$\frac{\partial E_x}{\partial z} = -\mu \frac{\partial H_y}{\partial t}.$$

This is the only component of curl E, so H_y is the only component of the magnetic field. Since E_x does not vary with x neither does H_y. Since div $H = 0$, H_y does not vary with y. Hence eqn (6.16) gives

(6.23)
$$\frac{\partial^2 H_y}{\partial z^2} = \mu\varepsilon \frac{\partial^2 H_y}{\partial t^2}.$$

Equations (6.21) and 6.23) are known as wave equations, because they have solutions of the form

(6.24a)
$$E = f_1(z - ct) + f_2(z + ct)$$

(6.24b)
$$H = F_1(z - ct) + F_2(z + ct)$$

where

(6.25)
$$c = \frac{1}{(\mu\varepsilon)^{1/2}}.$$

Examination of the wave equations shows that c has the dimensions of a velocity, and its numerical value is that of the velocity of light. This velocity is in the z-direction, that is, perpendicular to the electric and magnetic field vectors. The waves are transverse waves. Since light waves are also transverse, Maxwell inferred that light was an electromagnetic phenomenon.

The reason why the solutions in eqns (6.24a) and (6.24b) are called wave solutions can be clarified by reference to Fig. 6.1. Consider the function $f(z - ct)$ at two values of z and t. For the same value of the function

(6.26)
$$z_1 - ct_1 = z_2 - ct_2$$

or

(6.27)
$$z_2 - z_1 = c(t_2 - t_1)$$

Hence the entire function moves forward in the z-direction with velocity c. Similarly, a function $f(z + ct)$ moves backwards with velocity c.

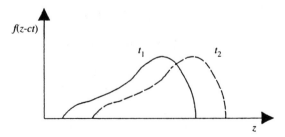

Fig. 6.1 A travelling wave.

E and *H* are related to each other through eqns (6.11) and (6.13). For the forward travelling component of the plane wave we have been considering,

$$(6.28) \qquad E_x = \left(\frac{\mu}{\varepsilon}\right)^{1/2} H_y.$$

For the backward travelling component,

$$(6.29) \qquad E_x = -\left(\frac{\mu}{\varepsilon}\right)^{1/2} H_y.$$

E and *H* are proportional to each other. In discussing static fields, magnetostatic and electrostatic effects could be separated; and in dealing with alternating currents in conductors the effect on the magnetic field due to the changing electric field could be ignored. But this separation of electricity and magnetism is not possible when dealing with electromagnetic waves. There is then a single electromagnetic entity which has electric and magnetic components.

Equations (6.28) and (6.29) were derived in the special simple case of a plane wave, but these equations are more general and they are also true for cylindrical and spherical waves. They derive from the symmetry of eqns (6.11) and (6.13) and apply where these equations provide the only sources of the field. This means that they apply at large distances from other sources such as conduction currents, polarizable materials, and magnetic materials. These equations describe the radiation of electromagnetic waves in free space.

Since *E* has units of V m^{-1} and *H* has units of A m^{-1} the combined wavefront carries a power of *EH* W m^{-2}. The direction is perpendicular to *E* and *H* and it can therefore be described by the vector

$$(6.30) \qquad \boldsymbol{S} = \boldsymbol{E} \times \boldsymbol{H}.$$

The positive direction of *S* is the direction of the forward-travelling wave. *S* is known as the Poynting vector after John Henry Poynting who first drew attention to it.

The energy per unit volume is

$$(6.31) \qquad U = \tfrac{1}{2}\varepsilon E^2 + \tfrac{1}{2}\mu H^2 = \varepsilon E^2 = \mu H^2,$$

since the two energy components are equal. Also,

$$(6.32) \qquad S = \left(\frac{\mu}{\varepsilon}\right)^{1/2} H^2 = cU.$$

Electromagnetic waves exert pressure on bodies which absorb or reflect them. Consider the pressure exerted by a plane wave at an absorbing surface. There is a compressive force perpendicular to the electric and magnetic field components, giving a pressure (in N m^{-2}) of

(6.33)
$$p = \tfrac{1}{2}\varepsilon E^2 + \tfrac{1}{2}\mu H^2.$$

In mechanics, force is equal to the rate of change of momentum. Consider pressure as the change of momentum as the electromagnetic wave is stopped by an absorbing surface. Let the wave have an equivalent mass density, ρ. Then during the small interval of time, dt, the mass (in kg m^{-2}) is given by

(6.34)
$$m = \rho c \, dt.$$

This mass has a momentum mc, and the rate of change of momentum is

(6.35)
$$\frac{d}{dt}(mc) = \rho c^2 = p.$$

Now

(6.36)
$$p = U,$$

as can be seen by comparing eqns (6.33) and (6.31). So

(6.37)
$$\rho c^2 = U.$$

Here ρ is mass/volume and U is energy/volume. Hence eqn (6.37) is a particular case of Einstein's famous equivalence relation between mass and energy

(6.38)
$$E = mc^2.$$

In general the radiation pressure is small. For example, in strong sunlight the power is of the order 1 kW m^{-2}. Hence the radiation pressure (in N m^{-2}) is

(6.39)
$$p = U = \frac{S}{c} = \frac{10^3}{3 \times 10^8} = \frac{1}{3 \times 10^5}.$$

The energy carried by an electromagnetic wave has to be supplied by the generator which launches that wave. Consider this in relation to a plane wave moving in the positive direction of z. To make the problem a little more realistic, consider an alternating time variation of angular frequency ω. Then

(6.40)
$$E = E_m e^{j(\omega t - \beta z)},$$

where the exponential combines the possibilities of $\sin(\omega t - \beta z)$ and $\cos(\omega t - \beta z)$. Also, $\omega t - \beta z$ must be a function of $z - ct$, so that β must be given by

(6.41)
$$\beta = \frac{\omega}{c}.$$

Similarly

(6.42)
$$H = H_m e^{j(\omega t - \beta z)}$$

and

(6.43)
$$E_m = \left(\frac{\mu}{\varepsilon}\right)^{1/2} H_m.$$

Close to the origin of coordinates,

(6.44)
$$H = H_m e^{j\omega t}.$$

Now consider a large flat current sheet at $z = 0$, as illustrated in Fig. 6.2. Let the current line density be $J_m e^{j\omega t}$ (A m^{-1}) in the x-direction. Close to the sheet, $\oint H \cdot dl = I$. Hence

(6.45)
$$2H_y = -J_m e^{j\omega t}.$$

Hence

(6.46)
$$H_y = -\tfrac{1}{2} J_m e^{j\omega t} = H_m e^{j\omega t}.$$

Hence

(6.47)
$$H_m = -\tfrac{1}{2} J_m$$

and

(6.48)
$$E_m = -\left(\frac{\mu}{\varepsilon}\right)^{1/2} \frac{J_m}{2}.$$

The average power needed to keep the current going is

(6.49)
$$\frac{1}{2} E_m J_m = \frac{1}{4} \left(\frac{\mu}{\varepsilon}\right)^{1/2} J_m^2.$$

The average power in the forward-moving wave is

(6.50)
$$S_z = \frac{1}{2} E_m H_m = \frac{1}{8} \left(\frac{\mu}{\varepsilon}\right)^{1/2} J_m^2.$$

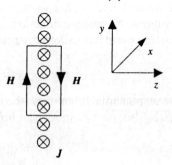

Fig. 6.2 A flat current sheet.

By symmetry, there is a wave, S_{-z}, on the other side of the sheet which has an equal power density. Hence the power supplied to the current exactly equals the power in the two waves being emitted by the sheet. Notice that the electric field at the sheet is in antiphase with the current. It is a *back e.m.f.* and it acts like a resistance, which is called the radiation resistance, R, where

(6.51)
$$R = -\frac{E}{J} = \frac{1}{2}\left(\frac{\mu}{\varepsilon}\right)^{1/2}.$$

The numerical value of R in free space is $188.4\ \Omega$. In an 'ohmic' resistance energy is converted to heat, and in the case of a radiation resistance energy is radiated into space. The numerical value of R depends on the configuration of the current (and charge) distribution on a radiating antenna.

6.3 Retarded potentials

The example of an infinite current sheet and plane waves of infinite area is of course highly idealized. It was chosen because its field is particularly simple inasmuch as the distant field and the field near the current have exactly the same form. In general, the near field is far more complicated than the distant field, and it is not possible to predict the near field from a knowledge of the distant field or vice versa.

For a connecting link, we turn to the potentials, which were useful in dealing with complicated, static, electric and magnetic fields. The static potentials cannot be used as they stand, because electromagnetic effects are propagated at the constant velocity of light. Hence space and time are linked by this velocity and cannot be treated independently. New expressions must, therefore, be developed for the potentials which take into account Maxwell's displacement currents. We have

(6.52)
$$\text{curl } \boldsymbol{H} = \boldsymbol{J} + \frac{\partial \boldsymbol{D}}{\partial t}.$$

Also, since

(6.53)
$$\text{div } \boldsymbol{B} = 0,$$

we can introduce the vector potential, A, through the relationship

(6.54)
$$\text{curl } \boldsymbol{A} = \boldsymbol{B}.$$

Now B and H are linked by the constitutive relation of the energy

(6.55)
$$\boldsymbol{B} = \mu \boldsymbol{H}.$$

Hence

(6.56) $$\text{curl } \boldsymbol{B} = \text{curl } \mu \boldsymbol{H} = \mu \text{ curl } \boldsymbol{H},$$

where μ has been taken as a constant. Hence

(6.57) $$\text{curl curl } \boldsymbol{A} = \mu \boldsymbol{J} + \mu \frac{\partial \boldsymbol{D}}{\partial t}.$$

From Faraday's law,

(6.58) $$\text{curl } \boldsymbol{E} = -\frac{\partial \boldsymbol{B}}{\partial t}.$$

Hence

(6.59) $$\text{curl } \boldsymbol{E} = -\text{curl } \frac{\partial \boldsymbol{A}}{\partial t}$$

and

(6.60) $$\boldsymbol{E} = -\frac{\partial \boldsymbol{A}}{\partial t} - \text{grad } V.$$

Notice that the 'uncurling' of eqn (6.59) introduces a constant of integration which is a gradient field.

The electric constitutive equation is

(6.61) $$\boldsymbol{D} = \varepsilon \boldsymbol{E}$$

so that

(6.62) $$\frac{\partial \boldsymbol{D}}{\partial t} = \varepsilon \frac{\partial \boldsymbol{E}}{\partial t} = -\varepsilon \frac{\partial^2 \boldsymbol{A}}{\partial t^2} - \varepsilon \text{ grad } \frac{\partial V}{\partial t}.$$

Substituting this in eqn (6.57),

(6.63) $$\text{curl curl } \boldsymbol{A} = \mu \boldsymbol{J} - \mu \varepsilon \frac{\partial^2 \boldsymbol{A}}{\partial t^2} - \mu \varepsilon \text{ grad } \frac{\partial V}{\partial t}.$$

Notice that there are now two terms involving the velocity of light. These have arisen from the displacement current term in eqn (6.52).

We can write

(6.64) $$\text{curl curl } \boldsymbol{A} = \text{grad div } \boldsymbol{A} - \nabla^2 \boldsymbol{A}$$

and insert this into eqn (6.63). Then

(6.65) $$\nabla^2 \boldsymbol{A} - \frac{1}{c^2} \frac{\partial^2 \boldsymbol{A}}{\partial t^2} = -\mu \boldsymbol{J} + \text{grad} \left(\text{div } \boldsymbol{A} + \frac{1}{c^2} \frac{\partial V}{\partial t} \right).$$

The left-hand side of this equation and the first term on the right-hand side give a wave equation similar to the waves previously considered. The gradient term complicates matters, but remember that A has so far been defined only by its curl, in eqn (6.54). Let us now define its divergence also in order to define it uniquely. Put

(6.66)
$$\text{div } A + \frac{1}{c^2} \frac{\partial V}{\partial t} = 0.$$

This provides a link between the vector and scalar potentials; it is called the Lorentz condition. Notice that in magnetostatics div $A = 0$ was used. This fits the new condition if there is no time variation. Now

(6.67)
$$\nabla^2 A - \frac{1}{c^2} \frac{\partial^2 A}{\partial t^2} = -\mu J.$$

In magnetostatics we had

(6.68)
$$\nabla^2 A = -\mu J,$$

which had the solution

(6.69)
$$A = \int_v \frac{\mu J}{4\pi r} \, dv.$$

The additional term in eqn (6.67) means that the space and time coordinates must always be associated through the velocity c. This can easily be done by writing

(6.70)
$$A_t = \int_v \frac{\mu [J]_{t'}}{4\pi r} \, dv$$

where $[J]_{t'}$ means that we must use the value of J at the earlier time $t' = t - r/c$, where t is the time at which A is observed. Notice that the earlier time depends on the position of the current J. Each piece of current contributes to the distant A in such a manner that its signal arrives at the same time as that of the other pieces of current. A_t is called the delayed or retarded vector potential.

Now consider V, the scalar potential. We have

(6.71)
$$\text{div } D = \rho,$$

where ρ is the charge density. Hence

(6.72)
$$\text{div } E = \frac{\rho}{\varepsilon},$$

where ε has been taken as a constant. From eqn (6.60)

(6.73)
$$-\frac{\partial}{\partial t} (\text{div } A) - \nabla^2 V = \frac{\rho}{\varepsilon},$$

and from eqn (6.66)

(6.74)
$$\nabla^2 V - \frac{1}{c^2} \frac{\partial^2 V}{\partial t^2} = -\frac{\rho}{\varepsilon},$$

so that

(6.75)
$$V_t = \int_v \frac{[\rho]_{t'}}{4\pi\varepsilon r} \, dv.$$

V_t is called the delayed or retarded scalar potential.

We now have a simple procedure for calculating the fields. $[J]$ gives A, and $[\rho]$ gives V. A gives the magnetic field, and A and V together give the electric field. We can, therefore, obtain the fields of any distribution of current and charge.

Notice one other interesting feature: current and charge are linked by the fact that currents are moving charges. This gives the continuity equation

(6.76)
$$\text{div } J + \frac{\partial \rho}{\partial t} = 0.$$

Compare this with the Lorentz condition of eqn (6.66). It is clear that the Lorentz condition has the same form as the continuity equation. This has removed the gradient term from eqn (6.65) and has ensured that A depends only on J, and that V depends only on ρ. Other choices for the divergence of A would give A in terms of J and ρ, and V similarly in terms of J and ρ. This would prevent the use of the simple delayed solution of eqns (6.70) and (6.75).

6.4 The field of an oscillating doublet or electric dipole

The smallest independent alternating electromagnetic system is a *current element* terminating in two charges, as illustrated in Fig. 6.3, and fed by a source of alternating current.

Previously, we considered a current element carrying a steady current

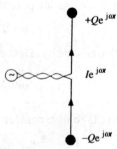

Fig. 6.3 An oscillating doublet.

and found that such an element had to be immersed in a conducting fluid. Such a fluid is unnecessary in the case of an alternating current terminated by charges. But the difference is not as clear-cut as might be thought, because the displacement current has introduced us to the idea that free space can carry current, although there is no conduction.

First let us calculate the magnetic field. This can be done by using the delayed vector potential. Consider this with reference to Fig. 6.4. A is parallel to the current element as shown by eqn (6.70). We have

(6.77)
$$A = \frac{\mu I \delta l}{4\pi r} e^{j(\omega t - \beta r)},$$

where r is very much greater than δl and I is taken to be constant over the length δl. Remember that $\beta = \omega/c$ and that it has the dimensions of (length)$^{-1}$. The characteristic length of a sine or cosine wave is the wavelength, λ, given by

(6.78)
$$\lambda = \frac{c}{f} = \frac{2\pi c}{\omega} = \frac{2\pi}{\beta}$$

so when r is equal to a multiple of λ, $e^{j\beta r}$ is $e^{j2n\pi}$, which is unity.

Since the vector potential depends on r, spherical coordinates (r, θ, ϕ) are used, as illustrated in Fig. 6.5. By symmetry, A does not depend on ϕ (the

Fig. 6.4 The magnetic field of a current element.

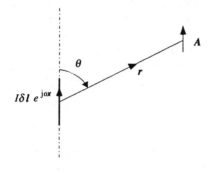

Fig. 6.5 Spherical coordinates.

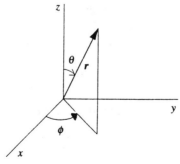

angle of longitude), nor does it have a component in the ϕ-direction. The two components are A_r and A_θ, where

(6.79)
$$A_r = A \cos \theta,$$

and

(6.80)
$$A_\theta = -A \sin \theta.$$

The magnetic field is given by $B = \text{curl } A$, and it has a single component, B_ϕ, (circles around the dipole), given by

(6.81)
$$B_\phi = \frac{1}{r} \frac{\partial}{\partial r} (r A_\theta) - \frac{1}{r} \frac{\partial A_\theta}{\partial \theta}.$$

This gives

(6.82)
$$B_\phi = \frac{\mu I \delta l e^{j(\omega t - \beta r)} \sin \theta}{4\pi r^2} (1 + \beta r).$$

This expression deserves careful examination. First, notice that as $\omega \to 0$, and therefore as $\beta \to 0$,

(6.83)
$$B_\phi \to \frac{\mu I \delta l \sin \theta}{4\pi r^2}.$$

This is the value for the Heaviside current element. The somewhat artificial idea of an element immersed in a conducting fluid of infinite extent is now validated as the limiting low-frequency case of an oscillating dipole. Secondly, notice that the magnetic field has two terms, one of which varies as $1/r^2$ and the other as β/r. Therefore, there are two limiting conditions, given by $\beta r \ll 1$ or $\beta r \gg 1$. Consider $\beta r \ll 1$. Then

(6.84)
$$B_\phi \approx \frac{\mu I \delta l e^{j\omega t} \sin \theta}{4\pi r^2},$$

where we have put $e^{-j\beta r} \approx 1$. Hence the magnetic field of an oscillating dipole for the condition $\beta r \ll 1$ can be obtained by taking the direct-current value and allowing it to alternate. The magnetic field is in phase with the current and it decays as $1/r^2$.

The condition $\beta r \gg 1$ leads to the expression

(6.85)
$$B_\phi \approx \frac{\beta \mu I \delta l e^{j(\omega t - \beta r)} \sin \theta}{4\pi r}.$$

In terms of the wavelength, λ, this is

(6.86)
$$B_\phi = \frac{\delta l}{\lambda} \frac{\mu I \sin \theta}{2r} e^{j(\omega t - \beta r)}.$$

This field has a phase varying with distance, and its magnitude decays as $1/r$. At large distances, therefore, the magnetic field is very much stronger than would have been expected from considering the steady-current field. The magnitude of the field depends on the ratio of the length δl to the wavelength λ.

In terms of λ, the two conditions on βr are $r \ll \lambda/2\pi$ or $r \gg \lambda/2\pi$. Magnitude can now be ascribed to the ideas of *near field* and *far field*. Moreover, eqn (6.84) shows that the near-field value is the value which would be obtained from inductance calculations based on Ampère's law without the displacement current. The far-field term arises from the delayed potential, which is itself based on the inclusion of displacement current. This term has a wave solution.

In terms of numerical results, $\lambda = c/f$, so that at 50 Hz $\lambda = 3 \times 10^8/50$ m $= 6 \times 10^6$ m. Hence for power-frequency calculations, $r \ll \lambda/2\pi$ is the dominant condition, and inductance calculations can be used. Radiation becomes important at frequencies for which $r \gg \lambda/2\pi$.

We can obtain the electric field for the radiation condition by noting that the radiation will be radially outward, so that

(6.87)
$$S_r = E_\theta H_\phi$$

and

(6.88)
$$E_\theta = \left(\frac{\mu}{\varepsilon}\right)^{1/2} H_\phi.$$

Hence

(6.89)
$$E_\theta = \left(\frac{\mu}{\varepsilon}\right)^{1/2} \frac{\delta l}{\lambda} \frac{I \sin \theta}{2r} e^{j(\omega t - \beta r)}.$$

The average radiated power is given by

(6.90)
$$P = \int_0^\pi \tfrac{1}{2} S 2\pi r^2 \sin \theta \, d\theta = \frac{\pi}{3} \left(\frac{\mu}{\varepsilon}\right)^{1/2} \left(\frac{\delta l}{\lambda}\right)^2 I^2.$$

The radiation resistance is given by

(6.91)
$$R = \frac{2P}{I^2} = \frac{2\pi}{3} \left(\frac{\mu}{\varepsilon}\right)^{1/2} \left(\frac{\delta l}{\lambda}\right)^2.$$

In free space $(\mu/\varepsilon)^{1/2} = 4\pi \times 10^{-7} \times 3 \times 10^8 = 120\pi$, so that the radiation resistance (in Ω) is

(6.92)
$$R = 80\pi^2 \left(\frac{\delta l}{\lambda}\right)^2.$$

The power must be supplied by the generator which causes the current to flow. Hence the electric field in antiphase to the current acting along the dipole is given by

(6.93)
$$E\delta l = -RI e^{j\omega t},$$

where E has been assumed to be uniform along δl. Hence

(6.94)
$$E = -80\pi^2 \frac{\delta l}{\lambda^2} I e^{j\omega t}.$$

It is instructive to check this value of the electric field by direct calculation from the potentials. The vector potential is given in eqn (6.77). The Lorentz condition relates the scalar potential to the vector potential by

(6.95)
$$\operatorname{div} A + \frac{1}{c^2} \frac{\partial V}{\partial t} = 0.$$

In spherical coordinates with r- and θ-components only

(6.96)
$$\frac{1}{r^2} \frac{\partial}{\partial r} (r^2 A_r) + \frac{1}{r \sin \theta} \frac{\partial}{\partial \theta} (A_\theta \sin \theta) = -\frac{j\omega}{c^2} V,$$

(6.97)
$$V = -j \frac{c^2}{\omega} \frac{\mu I \delta l \cos \theta}{4\pi r^2} e^{j(\omega t - \beta r)} (1 + j\beta r).$$

Now

$$E_r = -\frac{\partial A_r}{\partial t} - \frac{\partial V}{\partial r}$$

$$= -\frac{j\omega \mu I \delta l \cos \theta}{4\pi r} e^{j(\omega t - \beta r)}$$

(6.98)
$$+ j \frac{c^2}{\omega} \frac{\mu I \delta l \cos \theta}{4\pi} e^{j(\omega t - \beta r)} \left(-\frac{2}{r^3} - \frac{j\beta}{r^2} - \frac{j\beta}{r^2} + \frac{\beta^2}{r} \right).$$

Hence

(6.99)
$$E_r = \frac{jc\mu I \delta l \cos \theta}{2\pi \beta} e^{j(\omega t - \beta r)} \left(-\frac{1}{r^3} - j\frac{\beta}{r^2} \right).$$

We now expand $e^{-j\beta r}$ and find

(6.100)
$$E_r = \frac{jc\mu I \delta l \cos \theta}{2\pi \beta} e^{j\omega t} \left(1 - j\beta r - \frac{\beta^2 r^2}{2} + \frac{j\beta^3 r^3}{6} + \cdots \right) \left(-\frac{1}{r^3} - j\frac{\beta}{r^2} \right).$$

The dominant real and imaginary terms for the near field, where $\beta r \ll 1$, are

(6.101)
$$E_r = \frac{c\mu I \delta l \cos \theta}{2\pi \beta} e^{j\omega t} \left(-\frac{j}{r^3} - \frac{\beta^3}{3} \right).$$

There is a very large term in phase quadrature and a finite term in antiphase. The first gives the inductance and the second the radiation resistance. The inductance for a dipole of zero thickness would be infinite, but the power loss is finite because it depends on the radiated power, which is independent of the current density and only depends on the current.

Note that $\cos \theta = 1$ gives the electric field along the diple so that, for the finite antiphase component,

(6.102)
$$E = -\frac{c\mu I\delta l\beta^2}{6\pi} e^{j\omega t} = -80\pi^2 \frac{\delta l}{\lambda^2} Ie^{j\omega t},$$

which is the same as the value inferred in eqn (6.94) from consideration of the radiated power.

6.5 Radiation from a linear antenna

The short dipole discussed in the previous section can be used as the basic unit from which an arbitrary distribution of current can be obtained by superposition. As an example, consider the radiation from a half-wave linear antenna on which the current has a cosine distribution with a maximum at the centre and zero current at the ends, as illustrated in Fig. 6.6. The distribution can be written

(6.103)
$$i = I \cos \frac{2\pi z}{\lambda} = I \cos \beta z.$$

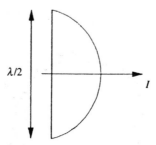

Fig. 6.6 A linear antenna.

Consider the radiation field at an angle θ, as illustrated in Fig. 6.7. The contribution from an element, δz, to the distant electric field, E_θ, is given by

(6.104)
$$\delta E_\theta = \left(\frac{\mu}{\varepsilon}\right)^{1/2} \frac{\delta z}{\lambda} I \cos(\beta z) \frac{\sin \theta}{2r_0} \exp[j(\omega t - \beta r_0 + \beta z \cos \theta)],$$

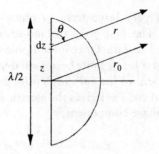

Fig. 6.7 A radiation field at an angle θ.

where the difference between r and r_0 is neglected in the denominator (but not in the numerator). Hence

(6.105)
$$E_\theta = \left(\frac{\mu}{\varepsilon}\right)^{1/2} \frac{I \sin \theta}{2\lambda r_0} \exp[j(\omega t - \beta r_0)] \int_{-\lambda/4}^{+\lambda/4} \exp(j\beta z \cos \theta)\cos \beta z \, dz,$$

whence

(6.106)
$$E_\theta = \frac{1}{2\pi}\left(\frac{\mu}{\varepsilon}\right)^{1/2} \frac{I}{r_0} \cos(\tfrac{1}{2}\pi \cos \theta)/(\sin \theta)e^{j(\omega t - \beta r_0)}.$$

In Fig. 6.8 the factor $\cos(\tfrac{1}{2}\pi \cos \theta)/\sin \theta$ has been plotted on a polar diagram, as has the factor $\sin \theta$ which occurs in the formula of the field of a dipole. This comparison shows that the half-wave antenna is somewhat

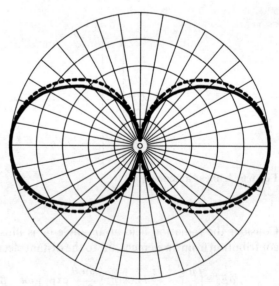

Fig. 6.8 The field of a linear antenna.

more 'directive'. The power is more concentrated in the directions near $\theta = \pi/2$.

6.6 Directivity of antenna arrays

We have seen that a half-wave antenna is more directive than a dipole. This directivity is due to the interaction of the waves radiated by the succession of dipoles placed end to end. This effect is even more marked when antennas are placed side by side. Consider two antennas perpendicular to the paper, as illustrated in Fig. 6.9. To simplify the treatment, consider the field in the horizontal plane. At a large distance ($\beta r \gg 1$) we have the combined electric field

$$E = E_0 \exp[j(\omega t - \beta r_1)] + E_0 \exp[j(\omega t - \beta r_2)]$$

$$= E_0 \exp[j(\omega t - \beta r_0)] \{\exp[j(\beta g/2)\sin \phi] + \exp[-j(\beta g/2)\sin \phi]\}$$

(6.107)
$$= 2E_0 \cos\left(\frac{\pi g}{\lambda} \sin \phi\right) \exp[j(\omega t - \beta r_0)],$$

where E_0 is the maximum value of the electric field of each member separately. Equation (6.107) should be compared with the field of the two members combined at the midpoint. This field is

(6.108)
$$E' = 2E_0 \exp[j(\omega t - \beta r_0)].$$

The factor $\cos[(\pi g/\lambda)\sin \phi]$ in eqn (6.107) reduces the field in certain directions, whereas the field of a single antenna is omnidirectional. Figure 6.10 shows the effect when the spacing between the two members is half a wavelength. For an array of N members spaced at $\lambda/2$ the reduction factor becomes

(6.109)
$$\frac{2}{N}\left[\cos\left(\frac{\pi}{2} \sin \phi\right) + \cos\left(\frac{3\pi}{2} \sin \phi\right) + \cdots + \cos\left(\frac{N-1}{2} \sin \phi\right)\right]$$

$$= \frac{\sin(\frac{1}{2}N\pi \sin \phi)}{N \sin(\frac{1}{2}\pi \sin \phi)}.$$

Fig. 6.9 Two parallel antennas.

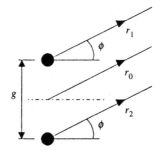

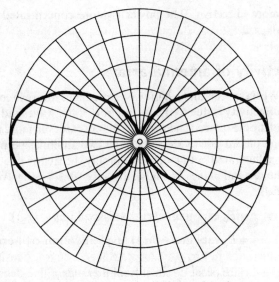

Fig. 6.10 Field of two parallel antennas a half wave-length apart.

The magnitude of this factor for an array of ten members is plotted in Fig. 6.11. The field now consists of a concentrated main beam in the forward and backward directions and a number of smaller side-lobes.

If a field in a certain direction is required it would be wasteful to use an omnidirectional antenna. An array of several members offers a saving in power by reducing this field in the unwanted directions. The *power gain* is defined as the power used by an omnidirectional antenna divided by the power used by a directional one.

Antenna design is a considerable subject in its own right. Our treatment has been superficial inasmuch as the two-dimensional radiation patterns in Fig. 6.10 and 6.11 do not represent the three-dimensional field adequately.

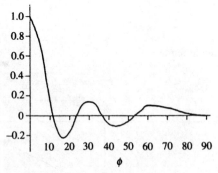

Fig. 6.11 A field plot for an array of 10 members.

Also, we have assumed that the currents in the members of an array are co-phased. This assumption has given a maximum field perpendicular to the plane of the array. With suitable adjustment, the main beam can be produced along the array. This is done in the Yagi arrays used for television.

Our purpose has been to illustrate the principle of the diffraction of electromagnetic waves, which is the cause of the directional property. In some directions the waves reinforce each other, and in other directions they cancel or reduce each other. The path difference in the waves caused by the spacing results in a phase difference. These differences depend on the angular direction of the distant point with respect to the antenna.

6.7 Waves guided by conductors

Consider a plane wave moving parallel to the surface of a conductor as illustrated in Fig. 6.12. There will be a charge density of $q = \varepsilon E_x$ C m^{-2} on the surface of the conductor, and a current line-density of $I_z = -H_y$ A m^{-1} in the surface layers of the conductor. From eqns (6.40) and (6.42)

$$E = \left(\frac{\mu}{\varepsilon}\right)^{1/2} H_m e^{j(\omega t - \beta z)}$$

and

$$H = H_m e^{j(\omega t - \beta z)}.$$

Hence

(6.110)
$$q = \frac{H_m}{c} e^{j(\omega t - \beta z)}$$

and

(6.111)
$$I = H_m e^{j(\omega t - \beta z)}.$$

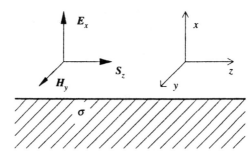

Fig. 6.12 A plane wave moving parallel to a conductor surface.

The continuity equation of charge and current is given by

$$\operatorname{div} \boldsymbol{J} + \frac{\partial \rho}{\partial t} = 0$$

which in our case gives

(6.112)
$$-j\beta H_m e^{j(\omega t - \beta z)} + j\frac{\omega}{c} H_m e^{j(\omega t - \beta z)} = 0,$$

which is correct. So the current and charge form a consistent system.

However, the current in the conductor needs an electric field to drive it and therefore an electric field is needed in the z-direction. Suppose that the conductor is thick, so that its equivalent resistance and reactance are $1/\sigma\Delta$, where Δ is the skin depth. Hence the magnitude of E_z will be $\sqrt{2}\,I/\sigma\Delta$. Hence the ratio of the magnitudes of E_x and E_z will be

(6.113)
$$\left|\frac{E_x}{E_z}\right| = \left(\frac{\mu}{\varepsilon}\right)^{1/2}\frac{\sigma\Delta}{2^{1/2}} = 267\sigma\Delta.$$

For copper at 50 Hz, $\Delta = 10^{-2}$ m, so that at 5 GHz $\Delta = 10^{-6}$ m. The conductivity of copper is of the order 10^7 S m^{-1}. So that at 5 GHz $\sigma\Delta$ is of the order of 10, and it is clear that even for such a high frequency E_z is negligibly small compared with E_x. It is therefore safe to assume that the electric field is perpendicular to the conductor. In terms of energy, the electric and magnetic field energies are very much larger than the resistive loss and the magnetic field energy in the conductor.

Now consider a plane wave between two conductors, as illustrated in Fig. 6.13. We now have a wave 'guided' by two conductors. The electric and magnetic field vectors are transverse to the direction of propagation and such a wave is called a TEM (transverse electric and magnetic) wave. The treatment can be generalized to deal with any two conductors supporting a TEM wave. For example, we could have two cylindrical wires as shown in

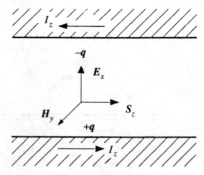

Fig. 6.13 A plane wave between two conductors.

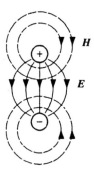

Fig. 6.14 A transmission line.

Fig. 6.14, where the direction of the electric field has been reversed, so that the top conductor has a positive charge and voltage. The direction of the wave is given by the Poynting vector $S = E \times H$ so that the wave is travelling into the paper in Fig. 6.14.

Clearly Fig. 6.14 shows a transmission line and the TEM mode is also called the transmission-line mode. An equivalent circuit diagram can now be drawn for TEM waves, as in Fig. 6.15. The electric energy is represented by a distributed capacitance and the magnetic field by a distributed inductance. The ohmic resistance has been omitted, because a transverse electric field is represented. The voltage drop across an element of inductance is given by

(6.114)
$$\delta V = L\, \delta z\, \frac{\mathrm{d}I}{\mathrm{d}t}$$

so that

(6.115)
$$\frac{\mathrm{d}V}{\mathrm{d}z} = L\,\frac{\mathrm{d}I}{\mathrm{d}t}.$$

The current across an element of capacitance is given by

(6.116)
$$\delta I = C\, \delta z\, \frac{\mathrm{d}V}{\mathrm{d}t}$$

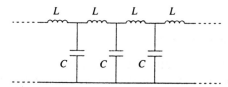

Fig. 6.15 An equivalent circuit for TEM waves.

so that

(6.117)
$$\frac{dI}{dz} = C\frac{dV}{dt}.$$

Hence

(6.118)
$$\frac{d^2V}{dz^2} = LC\frac{d^2V}{dt^2}$$

and

(6.119)
$$\frac{d^2I}{dz^2} = LC\frac{d^2I}{dt^2}.$$

But

(6.120)
$$LC = \mu\varepsilon = \frac{1}{c^2}$$

so that we have voltage and current waves of velocity c. As expected, the circuit representation of a loss-free transmission line leads to a TEM wave. It is interesting to note that it is even possible to produce a TEM wave at power frequencies by having a voltage high enough to make the electric field dominantly transverse to the conductors.

A similar result can be obtained with a cable as illustrated in Fig. 6.16. Here the TEM configuration is produced by a radial electric field and a circular magnetic field.

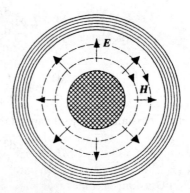

Fig. 6.16 A TEM wave in a cable.

6.8 Waveguide modes

A TEM wave requires two conductors which are insulated from each other. It would be useful if electromagnetic energy could be propagated inside tubular conductors and we shall show that this is indeed possible at high

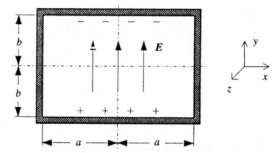

Fig. 6.17 A rectangular waveguide.

frequencies. Consider a rectangular tube as illustrated in Fig. 6.17, which shows the cross-section of the tube. Suppose we seek a propagating wave with its electric field in the y-direction. Such a field can be terminated by suitable charge distributions on the top and bottom of the tube. However, since we are dealing with a guide of negligible ohmic loss, E must be negligible or zero along the vertical sides. A suitable distribution would be

(6.121)
$$E_y = E_m \cos \frac{\pi x}{2a} e^{j(\omega t - \beta z)}.$$

To fit the wave equation we have to satisfy

(6.122)
$$\frac{\partial^2 E}{\partial x^2} + \frac{\partial^2 E}{\partial z^2} = \frac{1}{c^2} \frac{\partial^2 E}{\partial t^2}$$

noting that $\partial E/\partial y = 0$. Hence

(6.123)
$$-\frac{\pi^2}{4a^2} - \beta^2 = -\frac{\omega^2}{c^2}$$

or

(6.124)
$$\beta^2 = \frac{\omega^2}{c^2} - \frac{\pi^2}{4a^2} = \beta_0^2 - \frac{\pi^2}{4a^2},$$

where β_0 is the propagation constant of the TEM wave. The new propagation constant becomes imaginary if

(6.125)
$$\frac{\pi^2}{4a^2} > B_0^2$$

or

(6.126)
$$a^2 < \frac{\lambda^2}{16}.$$

An imaginary value for β implies that the wave would not propagate but would decay exponentially. There is therefore a *cut-off* if the width of the guide is less than half a wavelength. The cut-off frequency is given by

(6.127)
$$f \leq \frac{c}{4a}.$$

For a radar frequency of 10 GHz this means a width of more than 30 mm.

The magnetic field corresponding to the electric field of eqn (6.121) can be obtained from

$$\text{curl } \boldsymbol{E} = -\frac{\partial \boldsymbol{B}}{\partial t}.$$

This gives

(6.128)
$$H_x = -\frac{\beta}{\omega\mu} E_{\mathrm{m}} \cos \frac{\pi x}{2a} \, e^{j(\omega t - \beta z)}$$

and

(6.129)
$$H_z = -j \frac{\pi}{2a} \frac{E_{\mathrm{m}}}{\omega\mu} \sin \frac{\pi x}{2a} \, e^{j(\omega t - \beta z)}.$$

Notice that H_x is zero at the vertical sides, which is correct, because H has to be tangential to a perfect conductor. Notice also that E_y and H_x are in antiphase, so that the energy is conveyed in the direction $+z$. E_y and H_z are in phase quadrature so that no average energy is conveyed in the x-direction.

We have now established that a suitably wide tube will allow a wave to propagate with components E_y, H_x, and H_z. Since there is an H_z-component, this is not a TEM wave, but a transverse electric (TE) wave.

The reader should easily be able to show that there are other possible modes. TE modes can be constructed which have E_x- and E_y-components, and transverse magnetic (TM) modes can be constructed which have H_x- and H_y-components. Consideration of the boundary conditions for magnetic fields shows that there is no TM wave which has a single H_x- or H_y-component. In general, the boundary conditions can be met by using factors such as $\cos px$, $\sin px$, $\cos qy$ and $\sin qy$, where $p = m\pi/2a$ and $q = n\pi/2b$. The modes are classified as TE_{mn} and TM_{mn} modes. Thus the mode we have discussed is the TE_{10} mode. The cut-off conditions are given by

(6.130)
$$\frac{m^2}{4a^a} + \frac{n^2}{4b^2} = \frac{4}{\lambda^2}.$$

Exercises

6.1 Explain the term *displacement current*. A single-frequency plane wave is propagating in a medium for which $\sigma = 10 \text{ S m}^{-1}$, $\varepsilon = 1000\varepsilon_0$, and $\mu = \mu_0$. At what frequency will the conduction current and displacement current densities be equal?
(*Answer* 180 MHz.)

6.2 A cylindrical conductor of radius a carries a steady current I. Show that the inflow of the Poynting vector to the conductor is equal to the ohmic loss in the conductor.
(*Solution* Over the conductor surface we have $E = J/\sigma$ and $H = I/2\pi a$. Also $I = \pi a^2 J$. Hence $S = I^2/2\pi^2 a^3 \sigma$. Thus the power flow is

$$P = \oint S \cdot ds = \frac{I^2}{2\pi^2 a^3 \sigma} 2\pi a l = \frac{l}{\sigma \pi a^2} I^2 = RI^2.)$$

6.3 A polarized plane electromagnetic wave travels in a positive direction parallel to the z-axis. The electric field strength is $E_x = 1.2 \sin[(\omega/c)(z - ct)]$. Find an expression for the magnetic field strength.
(*Answer* $H_y = (1/100 \, \pi)\sin[(\omega/c)(z - ct)] \text{ A m}^{-1}$.)

6.4 The wave in Exercise 6.3 is generated by a current flowing in a thin plate which lies in the xy-plane at $z = 0$. Another wave of equal amplitude is emitted in the negative z-direction. Find an expression for this current.
(*Answer* $J_x = (1/50 \, \pi) \sin \omega t \text{ A m}^{-1}$.)

6.5 Find the radiation resistance per unit length and unit width of the thin plate in Exercise 6.4.
(*Answer* $60 \, \pi \, \Omega$.)

6.6 Explain why the delayed potentials A and V are dependent on each other. Show that they can be replaced by the single vector Π, called the Hertz vector, where

$$A = \frac{1}{c^2} \frac{\partial \Pi}{\partial t}, \qquad V = -\text{div } \Pi.$$

Show also that the magnetic and electric fields are given by

$$B = \frac{1}{c^2} \frac{\partial \text{ curl } \Pi}{\partial t}, \qquad E = \text{curl curl } \Pi.$$

6.7 In Section 6.4 the radiation from an electric dipole was analysed in terms of the potentials A and V. As an alternative, use the Hertz vector defined in Exercise 6.6 to show that

$$\Pi = \frac{Q\,\delta l}{4\pi\varepsilon_0 r}\,e^{j(\omega t - \beta r)},$$

and derive the electric and magnetic fields from the Hertz vector.

6.8 Determine the magnetic field of an oscillating electric dipole and show that it can be divided into a radiation component and an induction component. At what distance are the two components equal in magnitude? At what distance is the radiation field (a) 100 times the induction field, and (b) 1/100 times the induction field? Discuss these results in relation to the radiation from apparatus at power frequency. (*Answer* $r = c/\omega = \lambda/2\pi$, (a) $r = 15.9\lambda$, (b) $r = \lambda/629$, at 50 Hz $\lambda \approx 6000$ km.)

6.9 Ampère demonstrated the equivalence between a loop of constant current and a magnetic dipole as far as the *distant* field is concerned. What additional condition is needed to extend this equivalence to alternating currents?
(*Answer* The dimensions of the loop must be small compared to the electromagnetic wavelength.)

6.10 Explain what is meant by the *cut-off frequency* of a waveguide. Show that the lowest cut-off frequency for a TM mode is $\sqrt{2}$ times the lowest cut-off frequency for a TE mode if the cross section of the waveguide is a square.

Computation of fields

7.1 Introduction

So far in this book we have been chiefly concerned with the physical processes described by electromagnetic fields. Such computation as we have undertaken in the text, or encouraged the reader to undertake in the examples, has served primarily as an illustration of physical principles. This has meant that we have dealt only with very simple problems. In this chapter we transfer our attention to the computational procedures which enable engineers to find the field distributions in complicated pieces of apparatus and in electromagnetic systems. Our hope is that by now readers are familiar with the vocabulary of engineering electromagnetics and that after mastering the contents of this chapter they will be able to select and assess complicated computer packages as well as to write simple computer programs and to read specialized books and papers relevant to their particular interests.

There is an almost bewildering range of computational methods, each with its enthusiastic supporters. There are few people who are proficient in all the methods, just as there are very few engineers fully conversant with the whole range of electronic and power devices. It would be simply impossible to discuss all the available methods. What we shall attempt in this chapter is to provide a survey of the more commonly used computation schemes, highlighting the similarities and differences, and emphasizing the advantages and difficulties. We shall give a more detailed account of some of the most popular methods, such as *separation of variables*, *images*, *analogue methods*, *finite differences*, *finite elements*, *boundary elements*, and *tubes and slices*. The method of tubes and slices has already been used extensively in preceding chapters as a convenient platform for introducing basic concepts and illustrating the behaviour of fields. A copy of the tubes-and-slices (TAS) software is provided with the book, and Appendix 1 contains a tutorial on how to make full use of the program. Many other techniques, such as *conformal transformations*, *Laplace transforms*, or *transmission-line modelling*, will not be discussed in this book and the reader is referred to specialist texts. Some suggestions for further reading may be found in the Bibliography in Appendix 5.

Amongst the many available methods, an exact analytical solution is usually considered the most satisfactory. If such a solution may be found it will provide the designer with a lot of useful information in a convenient and compact form. But when an exact analytical solution is difficult or impossible to derive, we may resort to a numerical approximation or to an experimental or a graphical solution. The various methods of field modelling may be classified as follows:

(1) analogue methods;
(2) analytical solutions of the field equations;
(3) numerical methods (algebraical compution schemes); and
(4) graphical computation schemes.

The above subdivision should not be applied too rigorously; many practical techniques may span more than one category. For example, analytical solutions are often obtained in the form of infinite series or complicated integrals, so that computation necessitates the use of numerical techniques. Nevertheless, the classification helps to focus attention on the very essence of particular methods.

In this age of interactive graphics and with the enormous 'number crunching' capabilities of modern computers, it is not surprising to see numerical and graphical methods dominating the scene. The *computation of fields* should be seen as a broader aspect of *computer-aided design* in electromagnetics. It is also worth remembering that there is always more than one way of solving a particular problem and that understanding the solution is as important as knowing the numerical answer.

7.2 Separation of variables

The method of *separation of variables* is perhaps the most straightforward way of solving field equations and it is particularly useful for systems having geometrically simple boundaries. In this method, the appropriate partial differential equations are broken down into ordinary differential equations, and their solutions are combined to fit the boundary conditions of the problem. As an example, consider Laplace's equation in a rectangular system of coordinates (x, y, z) in two dimensions, that is, with no variation in the z-direction,

(7.1)
$$\nabla^2 \phi = \frac{\partial^2 \phi}{\partial x^2} + \frac{\partial^2 \phi}{\partial y^2} = 0.$$

Now assume that ϕ is expressible as the product of two quantities X and Y,

(7.2)
$$\phi = X(x) \, Y(y),$$

where X is a function of x alone and Y is a function of y alone.

Differentiation of eqn (7.2) and substitution in eqn (7.1) gives

(7.3)
$$\frac{1}{X}\frac{\partial^2 X}{\partial x^2} = -\frac{1}{Y}\frac{\partial^2 Y}{\partial y^2}.$$

Thus we have *separated* Laplace's equation into two parts, either term being a function of one variable only.

As x and y may be varied independently in any manner, and their sum must always be zero, the only way this can be achieved is by postulating that each term is equal to a constant, say $-p^2$, which is sometimes called the *separation constant*. We have made this particular choice of the constant because we seek sinusoidal or exponential functions. Then eqn (7.3) splits into two ordinary differential equations

(7.4)
$$\frac{d^2 X}{dx^2} + p^2 X = 0,$$

(7.5)
$$\frac{d^2 Y}{dy^2} - p^2 Y = 0.$$

These equations are satisfied by

(7.6)
$$X(x) \propto \begin{Bmatrix} \sin px \\ \cos px \end{Bmatrix}$$

(7.7)
$$Y(y) \propto \begin{Bmatrix} \sinh py \\ \cosh py \end{Bmatrix} \quad \text{or } e^{\pm py}.$$

The above result may be written in a more compact form as

(7.8)
$$\phi = \frac{\sinh}{\cosh}(py)\frac{\sin}{\cos}(px)$$

or

(7.9)
$$\phi = e^{\pm py}\frac{\sin}{\cos}(px).$$

The multiplication of each solution by an arbitrary constant will still give a solution of Laplace's equation, as does any sum of such terms. Thus, the final solution can be written by superposition as

$$\phi = \sum_{n=1}^{\infty} \{[K_n \sinh p_n y + L_n \cosh p_n y]\sin p_n x$$

(7.10)
$$+ [M_n \sinh p_n y + N_n \cosh p_n y]\cos p_n x\},$$

where K_n, L_n, M_n, N_n, and p_n are constants which need to be adjusted to fit the particular boundary conditions.

A special case arises if $p=0$, when

(7.11)
$$\frac{d^2X}{dx^2} = \frac{d^2Y}{dy^2} = 0$$

with solutions of the form

(7.12)
$$X(x) = Ax + B,$$

(7.13)
$$Y(y) = Cy + D,$$

giving an extra term

(7.14)
$$Exy + Fx + Gy + H$$

which may be added to the general solution of eqn (7.10) if necessary.

If ϕ is a function of all three space coordinates and Laplace's equation is

(7.15)
$$\nabla^2\phi = \frac{\partial^2\phi}{\partial x^2} + \frac{\partial^2\phi}{\partial y^2} + \frac{\partial^2\phi}{\partial z^2} = 0,$$

then letting

(7.16)
$$\phi = X(x)\,Y(y)\,Z(z),$$

then differentiation and substitution in eqn (7.15) gives

(7.17)
$$\frac{1}{X}\frac{\partial^2X}{\partial x^2} + \frac{1}{Y}\frac{\partial^2Y}{\partial y^2} + \frac{1}{Z}\frac{\partial^2Z}{\partial z^2} = 0.$$

Two arbitrary separation constants are needed as follows

(7.18)
$$\frac{d^2X}{dx^2} + p^2X = 0,$$

(7.19)
$$\frac{d^2Y}{dy^2} + q^2Y = 0,$$

(7.20)
$$\frac{d^2Z}{dz^2} - (p^2 + q^2)Z = 0.$$

In this way, Laplace's equation is separated into three ordinary differential equations. Thus the general solution may be expressed in the following compact form

(7.21)
$$\phi = \exp[\pm(p^2 + q^2)^{1/2}z]\,{\sin\atop\cos}\,(px)\,{\sin\atop\cos}\,(qy).$$

Combinations of such solutions can be built up to fit the boundary conditions.

The method will be understood more easily by considering the examples in the next two sections. Before leaving this section, however, let us make a general comment that while Laplace's equation yields the simplest formulation, the method can be applied to other field equations such as the diffusion and wave equations. The main limitation of the method is that it does not cope well with complicated boundaries so that only geometrically simple problems will usually be attempted. We shall see later in this chapter that graphical and numerical methods remove this restriction.

7.3 Separation of variables in electrostatics

Consider the potential distribution in a rectangular duct shown in Fig. 7.1. The potentials of the walls are specified so that a complete set of boundary conditions is available

(7.22)
$$\begin{cases} x=0, & V=0; \\ x=a, & V=0; \\ y=0, & V=f(x); \text{ and} \\ y=b, & V=0. \end{cases}$$

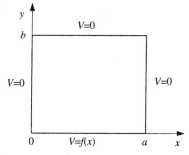

Fig. 7.1 The field in a rectangular region.

Consideration of the first boundary condition suggests that all coefficients M_n and N_n in eqn (7.10) must be zero, so that only terms containing $\sin p_n x$ may exist. At the same time, in order to satisfy the second condition for arbitrary values of K_n and L_n, we must have

(7.23)
$$\sin p_n a = 0.$$

Thus

(7.24)
$$p_n a = n\pi$$

or

(7.25)
$$p_n = \frac{n\pi}{a}, \quad \text{(where } n = 1, 2, 3, \ldots\text{).}$$

At this stage we can write

(7.26)
$$V = \sum_{n=1}^{\infty} \left(K_n \sinh \frac{n\pi y}{a} + L_n \cosh \frac{n\pi y}{a} \right) \sin \frac{n\pi x}{a}.$$

Investigation of the last of the boundary conditions for $y = b$ reveals that

(7.27)
$$K_n \sinh \frac{n\pi b}{a} + L_n \cosh \frac{n\pi b}{a} = 0.$$

After substitution into eqn (7.26) and some rearrangement we find

(7.28)
$$V = \sum_{n=1}^{\infty} L_n \frac{\sinh[(n\pi/a)(b-y)]}{\sinh(n\pi b/a)} \sin(n\pi x/a).$$

We now turn our attention to the third of the boundary conditions. When $y = 0$

(7.29)
$$V = \sum_{n=1}^{\infty} L_n \sin \frac{n\pi x}{a} = f(x).$$

Multiply both sides of eqn (7.29) by $\sin(m\pi x/a)$ and integrate from 0 to a

(7.30)
$$\sum_{n=1}^{\infty} \int_0^a L_n \sin \frac{n\pi x}{a} \sin \frac{m\pi x}{a} \, dx = \int_0^a f(x) \sin \frac{m\pi x}{a} \, dx.$$

The product of two sine terms integrates to zero over the given range (this is known as the orthogonality property) except when $n = m$, when

(7.31)
$$\int_0^a \sin^2 \frac{n\pi x}{a} \, dx = \int_0^a \frac{1}{2} \left(1 - \cos \frac{2n\pi x}{a} \right) dx = \frac{a}{2}.$$

Thus

(7.32)
$$L_n = \frac{2}{a} \int_0^a f(x) \sin \frac{n\pi x}{a} \, dx$$

and the problem is solved if $f(x)$ can be integrated. Moreover, eqn (7.32) can be recognized as the usual Fourier expansion of the function $f(x)$.

Take the simplest case of $f(x) = V_0$, that is, of a constant potential. Then

(7.33)
$$L_n = \frac{2V_0}{a} \int_0^a \sin \frac{n\pi x}{a} \, dx = \frac{2V_0}{a} \left[-\frac{a}{n\pi} \cos \frac{n\pi x}{a} \right]_0^a = \frac{2V_0}{n\pi} (1 - \cos n\pi).$$

If n is even

(7.34)
$$L_n = 0.$$

If n is odd

(7.35)
$$L_n = \frac{4V_0}{n\pi}.$$

Hence, by use of eqn (7.28),

(7.36)
$$V = \frac{4V_0}{\pi} \sum_{n=1,3,5,\ldots}^{\infty} \frac{1}{n} \frac{\sinh[(n\pi/a)(b-y)]}{\sinh(n\pi b/a)} \sin(n\pi x/a).$$

Specification of the potential on the boundary is known as the *Dirichlet boundary condition*. Mixed boundary conditions could also be introduced into the system. For example, the two vertical sides could have the so-called *Neumann condition* imposed by specifying the normal derivative of the potential, and thus, effectively, specifying the normal component of the electric field. A new set of boundary conditions applies:

(7.37)
$$\begin{cases} x=0, & \partial V/\partial x=0; \\ x=a, & \partial V/\partial x=0; \\ y=0, & V=f(x); \\ y=b, & V=0. \end{cases}$$

It can be easily shown that cosine rather than sine functions are present in the solution, but otherwise the derivation will be similar to the one presented before. This exercise is left to the reader.

The expressions found for the potential in the examples in this section satisfy the boundary conditions on the walls of the duct; this may be checked by simple substitution. Moreover these solutions describe the field *uniquely*. It has been shown before (in Section 4.9), but we will state again, that the potential within a closed region is uniquely specified by specification of the potential, or by specification of the normal electric field, at the boundary surface.

Finally, it is worth mentioning that for a more general problem, where a nonzero potential is specified on several sides, *superposition* offers the easiest solution. The idea is explained graphically in Fig. 7.2. Superposition is possible here because the system is linear. We must be careful, however, not to apply superposition to systems which are nonlinear. A nonlinear

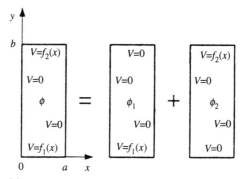

Fig. 7.2 Superposition.

system would contain implicit sources, which are dependent on the potential. For example the region might be filled with a material of varying permittivity.

7.4 Magnetostatic screening

Consider an infinitely long magnetic tube inserted into an applied uniform transverse field, H_0, as illustrated in Fig. 7.3. The tube has an external radius, a, an internal radius, b, and a constant permeability, μ_r. It will be convenient to solve the problem in terms of the *magnetic scalar potential*, defined by

(7.38)
$$H = -\operatorname{grad} V^*.$$

Moreover, because of the shape of the tube, a cylindrical coordinate system will be useful, where

(7.39)
$$\operatorname{grad} V^* = \hat{R}\frac{\partial V^*}{\partial R} + \hat{\phi}\frac{1}{R}\frac{\partial V^*}{\partial \phi} + \hat{z}\frac{\partial V^*}{\partial z}$$

and

(7.40)
$$\nabla^2 V^* = \frac{\partial^2 V^*}{\partial R^2} + \frac{1}{R}\frac{\partial V^*}{\partial R} + \frac{1}{R^2}\frac{\partial^2 V^*}{\partial \phi^2} + \frac{\partial^2 V^*}{\partial z^2}.$$

In our case, however, there will be no variation in the z-direction, because an infinitely long tube has been assumed. For the uniform applied field, H_0, acting in the x-direction

(7.41)
$$H_0\hat{x} = -\operatorname{grad} V_0^* = -\frac{\partial V_0^*}{\partial x}\hat{x}$$

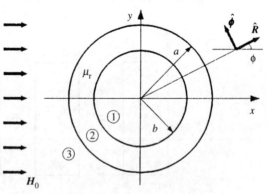

Fig. 7.3 The screening effect of an iron tube.

so that

(7.42)
$$V_0^* = -H_0 x = -H_0 R \cos \phi.$$

In the cylindrical coordinate system we assume that the general solution is in the form

(7.43)
$$V^* \propto R^{\pm n} \begin{Bmatrix} \sin n\phi \\ \cos n\phi \end{Bmatrix}$$

satisfying Laplace's equation in two dimensions. Three regions need to be considered in our case, but as the applied potential varies as $R \cos \phi$ we can deduce that we only require *cosine* solutions, and $n = 1$ in all three regions. In addition, outside the tube, where R becomes large, we cannot allow an $R \cos \phi$ term because the effect of the tube will become smaller as we move away from it. In region 1, the 'screened' region, we must discard $R^{-1} \cos \phi$, which would become infinite at $R = 0$. We therefore have

(7.44)
$$V_1^* = K_1 R \cos \phi,$$

(7.45)
$$V_2^* = (K_2 R + L_2 R^{-1}) \cos \phi,$$

(7.46)
$$V_3^* = (-H_0 R + L_3 R^{-1}) \cos \phi,$$

where the constants K_1, K_2, L_2, and L_3 can be found by using the four boundary conditions at the outer and inner radius. Thus, the continuity of the normal component of B and the tangential component of H across the two surfaces requires that

(7.47)
$$\frac{\partial V_1^*}{\partial R} = \mu_r \frac{\partial V_2^*}{\partial R} \quad \text{(for } R = b\text{)},$$

(7.48)
$$\frac{1}{b} \frac{\partial V_1^*}{\partial \phi} = \frac{1}{b} \frac{\partial V_2^*}{\partial \phi} \quad \text{or } V_1^* = V_2^* \quad \text{(for } R = b\text{)},$$

(7.49)
$$\mu_r \frac{\partial V_2^*}{\partial R} = \frac{\partial V_3^*}{\partial R} \quad \text{(for } R = a\text{)},$$

(7.50)
$$\frac{1}{a} \frac{\partial V_2^*}{\partial \phi} = \frac{1}{a} \frac{\partial V_3^*}{\partial \phi} \quad \text{or } V_2^* = V_3^* \quad \text{(for } R = a\text{)}.$$

This leads to the following set of equations

(7.51)
$$\begin{cases} K_1 = K_2 + L_2 b^{-2}, \\ K_1 = \mu_r (K_2 - L_2 b^{-2}), \\ K_2 + L_2 a^{-2} = -H_0 + L_3 a^{-2}, \\ \mu_r (K_2 - L_2 a^{-2}) = -H_0 - L_3 a^{-2}, \end{cases}$$

which after solving gives

(7.52)
$$\begin{cases} K_1 = -4H_0 a^2 \mu_r/D, \\ K_2 = -2H_0 a^2 (\mu_r + 1)/D, \\ L_2 = -2H_0 a^2 b^2 (\mu_r - 1)/D, \\ L_3 = a^2 H_0 + a^2 K_2 + L_2, \end{cases}$$

where

(7.53)
$$D = a^2(\mu_r + 1)^2 - b^2(\mu_r - 1)^2.$$

In region 1 inside the tube

(7.54)
$$H_{R1} = -\frac{\partial V_1^*}{\partial R} = -K_1 \cos \phi.$$

and

(7.55)
$$H_{\phi 1} = -\frac{1}{R}\frac{\partial V_1^*}{\partial \phi} = K_1 \sin \phi,$$

so that

(7.56)
$$\begin{cases} H_{x1} = -K_1 \\ H_{y1} = 0 \end{cases}$$

which implies that the field inside the tube is uniform. It will be helpful to introduce the *screening ratio*,

(7.57)
$$S = \frac{H_1}{H_0} = \frac{4\mu_r a^2}{a^2(\mu_r + 1)^2 - b^2(\mu_r - 1)^2}.$$

If $\mu_r \gg 1$,

(7.58)
$$S \approx \frac{4}{\mu_r}\frac{a^2}{a^2 - b^2} = \frac{4}{\mu_r}\frac{1}{1 - (b^2/a^2)},$$

and if the inside radius is small compared with the outside radius

(7.59)
$$S = \frac{4}{\mu_r}$$

so that the field in the screened region is $4/\mu_r$ times the original applied field. Finally, if the tube is thin,

(7.60)
$$S \approx \frac{1}{\mu_r}\frac{d}{t},$$

where d is the mean diameter and t is the thickness. However, the tube must be thick enough to avoid saturation, as the walls of the tube carry most of the flux. Otherwise the effective value of μ_r will drop rapidly and the screening action will be impaired.

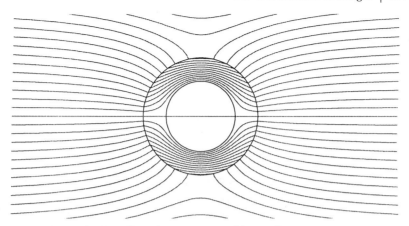

Fig. 7.4 The screening effect of an unsaturated iron tube.

A typical flux plot for the case of an unsaturated iron tube inserted into an originally uniform transverse field is shown in Fig. 7.4. The field inside the tube is very small and it can be seen that the screening is very effective. There is a large concentration of flux in the tube walls, and thus the flux density may exceed the saturation level.

7.5 The method of images

This useful method is based on analogies with the familiar phenomenon of images caused by the reflection of light in a mirror. There are electrostatic and magnetostatic images.

First, consider the electrostatic case. Figure 7.5(a) shows a point charge, q, in front of a conducting slab. There can be no static field in the slab. That means that the induced surface charge cancels the field of the point charge in the volume of the slab. The uniqueness theorem of Section 4.9 shows that the field in a closed volume is uniquely specified by the surface potential or its normal gradient plus the effect of the charge distribution within the volume. The field in the slab is therefore uniquely determined by the surface potential, which is constant and can be put to zero. The sources of this potential are the original charge and the induced surface charge. Hence, the surface charge cancels the effect of the original charge in the material, and it is equivalent to a negative point charge at the same place as the positive point charge.

To find the field above the slab we again need the zero surface potential and also the original charge. Figure 7.5(b) shows that the surface condition is correctly specified by an 'image' charge $-q$ at a distance d below the surface. Hence the field above the slab is that of the original charge and its

Fig. 7.5 The image of a point charge in a conducting slab.

image. Since complicated charge distributions can be obtained by superposition, the general case of a charge distribution above a flat conductor has been solved.

Next, consider the magnetostatic case as illustrated by Fig. 7.6(a) which shows a long current filament at a distance d above the surface of an iron slab of effectivly infinite permeability. Again, the surface sources cancel the effect of the applied current within the slab. The field above the slab is due to the surface polarity and the applied current. The contribution of the surface polarity is equal to that of the image current shown in Fig. 7.6(b). Notice that the image current has the same sign as the applied current in order to make the tangential field along the surface equal to zero and so produce a constant magnetostatic potential.

Now look at the more practical case of a finite permeability as in Fig. 7.7(a). This time the surface polarity does not cancel the applied field in the slab but it reduces it. From Section 3.3 (Equation 3.22) we know that the surface polarity is proportional to $(\mu_r - 1)/(\mu_r + 1)$ times the applied flux

Fig. 7.6 The image of a current filament in infinitely permeable iron.

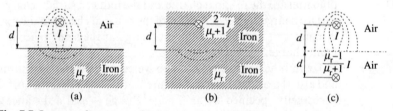

Fig. 7.7 Images of a current filament in iron of finite permeability.

density due to the current. Hence the effect in the iron is equivalent to a current source

$$I\left(1-\frac{\mu_r-1}{\mu_r+1}\right)=\frac{2I}{\mu_r+1}$$

as shown in Fig. 7.7(b). Above the slab, the surface polarity can be represented by an image current of strength

$$I\frac{\mu_r-1}{\mu_r+1}$$

as illustrated in Fig. 7.7(c) plus the original current.

The fields of the long current filament are two-dimensional, but the method can be applied to the three-dimensional fields of coils. Figure 7.8 illustrates the image currents which give the field above the iron. Inside the iron, the field will be that of the original coil multiplied by $2/(\mu_r+1)$.

The method can also be used when the boundary consists of several planes. Figure 7.9 shows a line current between two iron surfaces and a series of multiple images, which may be used for calculating the field in the air region. M is equal to $(\mu_r-1)/(\mu_r+1)$.

Another example involves intersecting boundaries. The range of such solutions is limited to a maximum of four boundaries. Also the angles of

Fig. 7.8 Images of coils of current in iron of finite permeability.

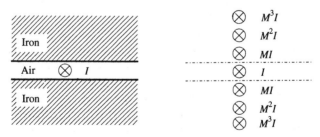

Fig. 7.9 Multiple images in two parallel iron plates.

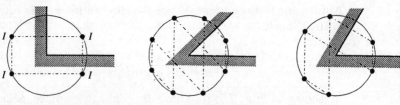

Fig. 7.10 Multiple images in intersecting boundaries.

interesection must be submultiples of π as can be seen by inspection of Fig. 7.10.

Finally we mention time-varying fields. These will induce eddy currents as discussed in Section 5.7. At very low frequencies and large values of μ_r, the surface polarity will be dominant over the eddy-current effect and approximate solutions can be obtained by the use of magnetostatic images. However, there will be a small phase lag in the image current due to the eddy-current loss. When the skin effect becomes dominant and the skin is very thin, the field cannot penetrate into the slab. This gives the so-called zero permeability condition, in which the external field can be obtained from an image of reversed sign. However, this approximation ignores the power loss. At frequencies such as those of light the electromagnetic field behaves as described in Chapter 6. The electric and magnetic field components are then proportional to each other, and the image becomes a true mirror image if the power loss is negligible. The sign of the image is then the same as that of electrostatic images and it is opposite to that of magnetostatic images.

7.6 Analogue methods

Analogue methods are based on an analogy between the mathematical description of various field systems. We have already identified and benefited from the analogy between electrostatic, magnetostatic, and steady-current-flow fields when introducing the method of *tubes and slices*, and we will come back to that discussion in the next section. In this section we shall explore some other possibilities offered by the principle of analogy. There are many useful experimental techniques for solving electromagnetic problems and we shall briefly discuss some of the methods.

Consider a pair of conducting electrodes in air and a pair of similar electrodes immersed in a conducting medium (with the additional proviso that the conductivity of the fluid is much lower than that of the electrodes). If a potential difference is applied between the electrodes of the former system, an electric field will be set up in which the conductor surfaces are equipotentials. For the latter system, application of a potential difference will set up a flow of steady current; but the two field maps are identical in terms of the distribution of potential surfaces (slices) and in the distribution of flux or electric-current boundaries (tubes). This principle may be used to investigate characteristics of the electric field between given electrodes, by making measurements on model electrodes immersed in a conducting fluid or electrolyte. The apparatus is known as an *electrolytic tank*. The methods used in electrolytic tanks have been extensively studied and refined in the past, and accuracies of a quarter of a per cent in measurement of potentials have been reported.

If the field is two-dimensional, the model can be set up in a shallow dish. Alternatively a graphitized paper may be used, where electrodes (equipotentials) of arbitrary shape can be painted on the paper with silver paint. A potential difference is applied between these electrodes, and the equipotentials may be easily drawn on the paper by moving a probe across the paper.

The magnetic-circuit analogy is realized experimentally when the field in the air space between iron surfaces is plotted by means of an electrolytic tank or conducting sheet, in a manner similar to that described for electric fields. The iron surfaces are then represented by metal electrodes of similar shape to the iron; this implicitly assumes that the permeability of the iron is infinite. For unsaturated iron this assumption is usually acceptable.

Another interesting analogue is provided by a mesh built from resistances, as illustrated in Fig. 7.11. Such a *network analogue* could be considered as a discretized form of the 'distributed' analogues of the electrolytic tank or the conducting sheet. Three-dimensional networks can also be built. The accuracy depends strongly on the mesh size, and in this respect network analogues resemble some of the numerical methods described later in this chapter. The use of a network avoids some of the difficulties of measurement in the other analogues, particularly those due to chemical effects. However, the real advantage arises when other circuit components are introduced, and thus other types of fields involving time variation may be modelled. This aspect of network analogues will not be

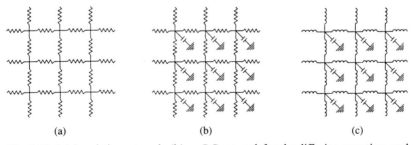

(a)   (b)   (c)

Fig. 7.11 (a) A resistive network, (b) an *RC* network for the diffusion equation, and (c) an *LC* network for the wave equation.

discussed here in any detail but appropriate circuits for other equation types are shown in Fig. 7.11.

Finally, we shall mention an interesting technique which requires very modest experimental equipment in the form of a sheet of paper, a pencil, and an eraser. It is essentially a graphical method based on subdivision of the entire region under consideration into curvilinear squares. The term *curvilinear square* means a right-angled figure that a process of regular subdivision will ultimately reduce to a collection of actual squares. With this modest apparatus, plus considerable patience, one can solve Laplace's equation for any two-dimensional case.

Consider Fig. 7.12 which shows some lines in an electrostatic field. We can write

$$-\delta V = E \, \delta x, \tag{7.61}$$

$$\delta\phi = D \, \delta y = \varepsilon E \, \delta y, \tag{7.62}$$

where the flux ϕ is taken per unit depth into the paper. If we choose $\delta x = \delta y$, because the eye easily detects differences between δx and δy, we have

$$\left|\frac{\delta\phi}{\delta V}\right| = \varepsilon. \tag{7.63}$$

Thus the total flux or the total potential difference can be found by counting the number of the flux lines or the equipotentials. The procedure of solving Laplace's equation by this graphical method starts with sketching a few equipotential and flux lines on a scale drawing of the region, then subdividing the resulting curvilinear squares until the required accuracy is obtained. In this mapping of the plane region, two criteria must be satisfied at each step:

1. Every intersection must be orthogonal.
2. Every subregion must appear to be a curvilinear *square* and not a curvilinear rectangle.

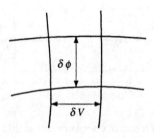

Fig. 7.12 Curvilinear squares.

As the mesh becomes finer, many lines may have to be redrawn to meet the above requirements. Thus a considerable proportion of the work consists of erasing of what has been drawn previously. When a finely divided map is obtained, however, satisfying conditions 1 and 2 throughout, it represents the unique solution of the problem. In experienced hands this is a quick and powerful method.

We have deliberately introduced the method of curvilinear squares as the last method in this section, because it provides a useful bridge to the method of *tubes and slices*. Indeed, the two methods have a lot in common, as they are both based on a geometrical approach, rather than an analytical or numerical formulation. In fact, no knowledge of mathematics is necessary to apply either of the methods successfully. They are also similar in that they use directly the field map as a means of finding the solution.

7.7 Tubes and slices

The method of *tubes and slices* (TAS) is advocated throughout this book as a method which is physically well based and computationally attractive. We have already discussed in great detail the application of this method to various electrostatic, magnetostatic, and steady-current-flow problems. A copy of the program accompanies this book and the reader is strongly encouraged to experiment and explore various features offered by the software. Appendix 1 contains a tutorial, and it is recommended that this is used for familiarization with the program and to learn how to take full advantage of its capabilities.

It is worth mentioning that the tubes-and-slices approach is essentially a variational technique which is implemented using geometrical, rather than algebraic, consideration. The short analysis which follows assumes some basic knowledge of functional analysis and it is included here for completeness, but it is not essential for understanding or using the method.

In magnetostatic fields the equilibrium conditions of the system can be described by two variational principles applied within the volume

(7.64) $$\langle (\nabla \times \boldsymbol{H} - \boldsymbol{\bar{J}}), \delta A \rangle = 0$$

and

(7.65) $$\langle (\nabla \times \boldsymbol{A} - \boldsymbol{B}), \delta H \rangle = 0,$$

where the angle brackets, $\langle \, \rangle$, indicate integration through the region of interest, and $\boldsymbol{\bar{J}}$ is the assigned current density.

The first variational principle assumed that $\boldsymbol{B} = \text{curl } \boldsymbol{A}$, so that div $\boldsymbol{B} = 0$. This means that there are no divergence sources for the magnetic field. However, the expression $\nabla \times \boldsymbol{H} - \boldsymbol{\bar{J}}$ allows a small variation in $\nabla \times \boldsymbol{H}$ from its correct value, so that the variation allows a small additional distribution of *curl* sources. The product of this small fictitious current multiplied by the small variation of A gives an energy variation of the second order (of small quantities) which can be put to zero.

The second variational principle assumes that curl $H = \bar{J}$, so that the *curl* sources of the magnetic field are correct. However, the expression $\nabla \times A - B$ allows a small variation in the *divergence* sources. The product of this small polarity distribution multiplied by the small variation of H gives an energy variation of the second order (of small quantities), which can be put to zero.

The field energy can be expressed either in terms of the field vectors H and B by

(7.66)
$$U = \tfrac{1}{2}\langle B, H \rangle,$$

or in terms of the interaction of the current sources with the vector potential, A, by

(7.67)
$$U = \tfrac{1}{2}\langle \bar{J}, A \rangle + \tfrac{1}{2}[\bar{I}, A],$$

where $\bar{I}$ is the assigned line density of current on the surface, and the square brackets, [ ] represent integration over the closed boundary surface. $\bar{I}$ is related to $\bar{J}$ in such a way as to make the total current in the system zero. This isolates the system and gives it a unique energy.

The first variational principle is applied to the energy in terms of A and $B = \text{curl } A$ by writing

(7.68)
$$\delta U(A) = \delta\{\langle \bar{J}, A \rangle + [\bar{I}, A] - \tfrac{1}{2}\langle B, B/\mu \rangle\} = 0.$$

The second variation is therefore negative, that is

(7.69)
$$\delta^2 U(A) \le 0.$$

The second variational principle is applied to the energy in terms of H by writing

(7.70)
$$\delta U(H) = \delta\{\tfrac{1}{2}\langle H, \mu H \rangle\} = 0.$$

Hence

(7.71)
$$\delta^2 U(H) \ge 0.$$

For simplicity, μ has been assumed to be constant, and this gives the factor $\tfrac{1}{2}$. However, this method is applicable to all permeabilities, which are single-valued functions of the field-strength.

The second variations show the possibility of obtaining both upper and lower bounds for the energy. The first variational principle treats the field as a collection of tubes and the second one as a collection of slices.

Many of the popular methods of field calculation do not distinguish between the two kinds of sources and so do not obtain bounded solutions. Moreover, the use of a free topology in discretization means that the local behaviour of the field is distorted, so that the fictitious sources have to be minimized for the entire system by means of a set of simultaneous equations, which may be a costly process.

This process is avoided by the use of *tubes and slices*, because each tube or slice can be treated as a separate entity, so that local variations and improvements can be made without having to involve the entire system. The program enclosed with this book is designed to solve simple problems in electrostatics, magnetostatics, and steady-current flow. Extension of this method to time-varying problems is possible, but this is beyond the scope of this book.

7.8 The finite-difference method

The *finite-difference* (FD) method has been a very successful numerical technique, leading to many general-purpose computer codes in electromagnetics. Although overshadowed nowadays by the more versatile finite-element method, discussed later in this chapter, it continues to play an important role in numerical analysis. One of the significant advantages of the FD method is the simplicity of its formulation, both in a mathematical and a numerical sense. A basic FD scheme could almost be set up intuitively, and a simple computer program for a particular solution could be written even by an inexperienced programmer in a matter of minutes. Moreover, the whole process is easily understood. However, there are some important difficulties encountered when using FD equations, and they will be addressed at the end of this section.

Consider Laplace's equation in two dimensions (eqn 7.1) in some arbitrarily shaped region with the boundary conditions specified. We will overlay the problem with a mesh of lines 'parallel' to the coordinate system used (a rectangular grid is convenient, though not essential), and we will seek an approximate solution at mesh points defined by intersections of the lines. With reference to Fig. 7.13, we consider the mesh point (i, j) and its

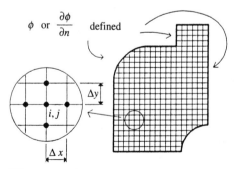

Fig. 7.13 A finite-difference mesh.

immediate neighbours, and by using Taylor's series we can write

(7.72)
$$\phi_{i+1,j} = \phi_{i,j} + \Delta x \left.\frac{\partial \phi}{\partial x}\right|_{i,j} + \frac{(\Delta x)^2}{2!} \left.\frac{\partial^2 \phi}{\partial x^2}\right|_{i,j} + \frac{(\Delta x)^3}{3!} \left.\frac{\partial^3 \phi}{\partial x^3}\right|_{i,j} + \cdots$$

and

(7.73)
$$\phi_{i-1,j} = \phi_{i,j} - \Delta x \left.\frac{\partial \phi}{\partial x}\right|_{i,j} + \frac{(\Delta x)^2}{2!} \left.\frac{\partial^2 \phi}{\partial x^2}\right|_{i,j} - \frac{(\Delta x)^3}{3!} \left.\frac{\partial^3 \phi}{\partial x^3}\right|_{i,j} + \cdots.$$

Adding the last two equations yields

(7.74)
$$\phi_{i+1,j} + \phi_{i-1,j} = 2\phi_{i,j} + (\Delta x)^2 \left.\frac{\partial^2 \phi}{\partial x^2}\right|_{i,j} + O((\Delta x)^4),$$

where $O((\Delta x)^4)$ represents terms containing fourth and higher-order powers of Δx. Neglecting these terms we have

(7.75)
$$\left.\frac{\partial^2 \phi}{\partial x^2}\right|_{i,j} = \frac{\phi_{i+1,j} - 2\phi_{i,j} + \phi_{i-1,j}}{(\Delta x)^2}.$$

Similarly, under the same assumptions,

(7.76)
$$\left.\frac{\partial^2 \phi}{\partial y^2}\right|_{i,j} = \frac{\phi_{i,j+1} - 2\phi_{i,j} + \phi_{i,j-1}}{(\Delta y)^2}.$$

If these FD expressions are now substituted into Laplace's equation, eqn (7.1), the following local approximation is obtained

(7.77)
$$\frac{1}{(\Delta x)^2}(\phi_{i+1,j} - 2\phi_{i,j} + \phi_{i-1,j}) + \frac{1}{(\Delta y)^2}(\phi_{i,j+1} - 2\phi_{i,j} + \phi_{i,j-1}) = 0.$$

If for convenience we choose a square mesh, so that $\Delta x = \Delta y$,

(7.78)
$$\phi_{i-1,j} + \phi_{i+1,j} + \phi_{i,j-1} + \phi_{i,j+1} - 4\phi_{i,j} = 0$$

or

(7.79)
$$\phi_{i,j} = \tfrac{1}{4}(\phi_{i-1,j} + \phi_{i+1,j} + \phi_{i,j-1} + \phi_{i,j+1}),$$

which is a well-known five-point scheme for Laplace's equation. The graphical representation is shown in Fig. 7.14. It is interesting to note that in this scheme the value at any node is taken as the average of the four values of its immediate neighbours. In this sense the scheme could be considered as intuitive, but we have demonstrated that Laplace's equation is in fact satisfied, subject to errors introduced by neglecting higher-order terms in the Taylor's series. These errors are due to the finite mesh size, and they are called *discretization* or *truncation* errors.

Equation (7.79) suggests a possible simple scheme for obtaining the solution by 'scanning' all nodes iteratively. Boundary nodes must have

Fig. 7.14 A five-point
computation scheme.

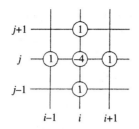

appropriate boundary conditions assigned, and then the information about outside sources of the field, as provided by the boundary conditions, will gradually spread to the interior of the region through successive application of eqn (7.79). Other methods will be discussed later in this chapter.

Consider the tubular capacitor illustrated in Fig. 7.15(a). Symmetry allows us to consider one quarter of the system only, and symmetry lines (planes) will have a boundary condition on $\partial V/\partial n$ (Fig. 7.15(b)). A very simple FD grid is shown in Fig. 7.15(c). Notice that the condition on $\partial V/\partial n$ can be easily satisfied by adding 'image' nodes to the grid, as indicated.

We apply the five-point formula (eqn 7.78) to the five nodes with unknown potentials, V_1, V_2, V_3, V_4, and V_5

(7.80)
$$\begin{cases} V_2 + 100 + V_2 + 0 - 4V_1 = 0, \\ V_1 + 100 + V_3 + 0 - 4V_2 = 0, \\ V_2 + 100 + V_4 + 0 - 4V_3 = 0, \\ V_3 + 100 + 100 + V_5 - 4V_4 = 0, \\ 0 + V_4 + 100 + V_4 - 4V_5 = 0. \end{cases}$$

This can be written in matrix form as follows:

(7.81)
$$\begin{bmatrix} 4 & -2 & 0 & 0 & 0 \\ -1 & 4 & -1 & 0 & 0 \\ 0 & -1 & 4 & -1 & 0 \\ 0 & 0 & -1 & 4 & -1 \\ 0 & 0 & 0 & -2 & 4 \end{bmatrix} \begin{bmatrix} V_1 \\ V_2 \\ V_3 \\ V_4 \\ V_5 \end{bmatrix} = \begin{bmatrix} 100 \\ 100 \\ 100 \\ 200 \\ 100 \end{bmatrix}.$$

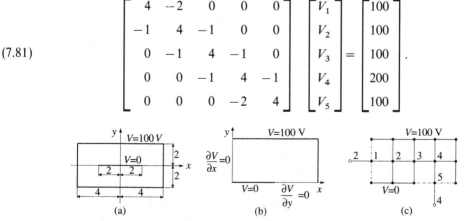

(a) (b) (c)

Fig. 7.15 A simple tubular-capacitor problem to illustrate the finite-difference procedure.

Equation (7.81) is in the standard form for a set of linear algebraic equations,

(7.82)
$$Ax = b,$$

and it may be solved using methods described briefly in Section 7.11 later in this chapter. In our case the solution could be easily obtained by hand, which is hardly worth the effort because the mesh used is very coarse and thus the solution will not be very accurate. It is more important to note that a finer mesh would lead to similar equations, but a great many of them. A simple computer program could be written to perform the task of assembling and then solving the system of equations. Figure 7.16 shows a typical refined FD mesh.

To summarize, we have seen how the FD scheme transforms a continuous partial differential equation to a discrete set of algebraic equations. The system matrix, A, corresponding to the differential operator (∇^2 in this case), is a square matrix, which means that the number of unknowns is equal to the number of equations. If a solution exists then the inverse of the matrix A must exist. In these aspects the FD method resembles the finite-element method which is described in Sections 7.9 and 7.10.

The FD method can readily be applied to the other differential equations met in electromagnetism. For example, the diffusion equation is of the form

(7.83)
$$\frac{\partial^2 \phi}{\partial x^2} = \alpha \frac{\partial \phi}{\partial t}$$

and the wave equation has the form

(7.84)
$$\frac{\partial^2 \phi}{\partial x^2} = \beta \frac{\partial^2 \phi}{\partial t^2}.$$

The time derivatives can be obtained with the aid of Taylor's series:

(7.85)
$$\frac{\partial \phi}{\partial t} = \frac{\phi_{i,k+1} - \phi_{i,k}}{\Delta t},$$

Fig. 7.16 A refined mesh.

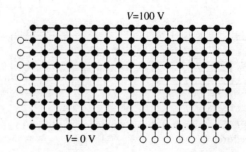

(7.86)
$$\frac{\partial^2 \phi}{\partial t^2} = \frac{\phi_{i,k+1} - 2\phi_{i,k} + \phi_{i,k-1}}{(\Delta t)^2},$$

where Δt is the time interval between successive values of ϕ_i appearing at the space node i, and the suffix k denotes the time variable. It may be helpful to think of the solution as marching forward in time, each step progressing by Δt, generating the electromagnetic transient as it goes.

Hence it follows that

(7.87)
$$\frac{\phi_{i+1,k} - 2\phi_{i,k} + \phi_{i-1,k}}{(\Delta x)^2} = \alpha \frac{\phi_{i,k+1} - \phi_{i,k}}{\Delta t}$$

for the diffusion equation, and

(7.88)
$$\frac{\phi_{i+1,k} - 2\phi_{i,k} + \phi_{i-1,k}}{(\Delta x)^2} = \beta \frac{\phi_{i,k+1} - 2\phi_{i,k} + \phi_{i,k-1}}{(\Delta t)^2}$$

for the wave equation, where i denotes a space variable and Δx is the distance between adjacent nodes. Examination of the last two equations reveals that we can obtain one point ahead in time from the values of the previous time row (for the diffusion equation), or from two previous time rows (for the wave equation). Assume that the boundary conditions are such that $\phi(x, t)$ is given for all t at $x=0$ and $x=L$, and that the initial condition for $\phi(x, 0)$ is specified for all x. Suitable computation schemes are shown in Figs 7.17 and 7.18. Both solutions move forward in time, row by row. This is known as an *explicit* FD scheme. Other formulations are also possible, where the solution at each time row is not calculated explicitly, but more than one unknown value is related to several known values in one equation. Such *implicit* schemes are often solved by iteration. Note also that special equations are required for nodes positioned next to the boundaries, even in the explicit scheme. Finally, any scheme must be *compatible* with the differential equation and it must be *stable*. Compatibility ensures that the numerical solution converges to the solution of the original equation. Stability may be lost if the various errors accumulate without limit during

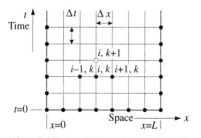

Fig. 7.17 An FD scheme for the diffusion equation.

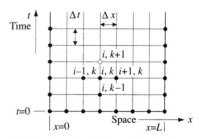

Fig. 7.18 An FD scheme for the wave equation.

the computation. Stability conditions can be derived for some equations. Thus for eqn (7.87) stability is obtained only if

(7.89)
$$\frac{\Delta t}{\alpha(\Delta x)^2} \le \frac{1}{2}$$

and for eqn (7.88) if

(7.90)
$$\frac{1}{\beta}\left(\frac{\Delta t}{\Delta x}\right)^2 \le 1.$$

Compatibility and stability may not always be easy to ensure and they are important issues, but a more detailed discussion is beyond the scope of this book.

A major difficulty with the FD method is due to the fixed topology (both order and arrangement) of the discretization scheme. It becomes very difficult to match highly irregular boundaries with an appropriate mesh or grid. At the same time, material interfaces, symmetry conditions, and nonlinear characteristics all require special treatment. Another difficulty arises when higher order terms from Taylor's series are to be introduced to improve the accuracy. Although several special algorithms have been developed, they do not offer as much versatility as the *finite-element method*.

7.9 The finite-element method

The *finite element* method has, in recent years, become by far the most popular technique in computational electromagnetics. Many general purpose computer codes have been developed which provide the basis for computer-aided-design (CAD) systems. The technique is not suitable for hand calculations, and the algorithm is somewhat complicated; nevertheless we shall attempt to follow the formulation of the simplest two-dimensional case to demonstrate the principle and we shall discuss some aspects of the applications. There is a vast literature on the subject, and the reader is encouraged to consult specialist books and papers to explore the more advanced features of the method.

Consider Laplace's equation in two dimensions for an electrostatic system

(7.91)
$$\nabla^2 V = \frac{\partial^2 V}{\partial x^2} + \frac{\partial^2 V}{\partial y^2} = 0,$$

where the electric field, E, is given by

(7.92)
$$E = -\operatorname{grad} V = -\nabla V = -\left(\frac{\partial V}{\partial x}\hat{x} + \frac{\partial V}{\partial y}\hat{y}\right).$$

We can apply a variational principle to the defining equation by stating that the principle of equilibrium requires that the potential distribution must be such as to minimize the stored field energy. In our case, this energy can be expressed as

$$W = \int_\Omega \tfrac{1}{2} \boldsymbol{E} \cdot \boldsymbol{D} \, d\Omega = \tfrac{1}{2} \int_\Omega \varepsilon E^2 \, d\Omega$$

(7.93)
$$= \frac{1}{2} \int_\Omega \varepsilon \left\{ \left(\frac{\partial V}{\partial x} \right)^2 + \left(\frac{\partial V}{\partial y} \right)^2 \right\} d\Omega = \tfrac{1}{2} \int_\Omega \varepsilon \, |\nabla V|^2 \, d\Omega,$$

where integration is carried out over the two-dimensional problem region, and is thus taken per unit length. This minimum-energy principle is mathematically equivalent to our original differential equation in the sense that a potential distribution which satisfies Laplace's equation will also minimize the energy, and vice versa.

At this stage we mention that an alternative formulation is possible which avoids using energy functionals. It is based on the so-called Galerkin procedure and the method of weighted residuals. The Galerkin procedure is generally easier to apply and leads to a wider class of applications. However, its mathematical formulation is more advanced and it will not be pursued here. For Laplace's equation both formulations give identical results.

Consider a single element, and assume that the potential distribution within the element is adequately approximated by the expression

(7.94)
$$V = a + bx + cy + dxy + ex^2 + fy^2 + \cdots .$$

We choose as many terms in the above equation as there are 'nodes' in the element. Figure 7.19 shows some examples. For a rectangle we choose

(7.95)
$$V = a + bx + cy + dxy$$

and for the first-order triangle

(7.96)
$$V = a + bx + cy.$$

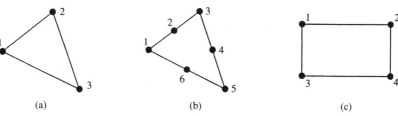

Fig. 7.19 Finite elements, (a) a first-order triangle, (b) a second-order triangle, and (c) a rectangle.

In the last case the representation is said to be complete because eqn (7.96) contains all the terms necessary for a linear variation in two dimensions. We shall not pursue the higher-order elements, but it is easily seen that finite elements, unlike finite differences, offer a very natural extension to higher-order modelling.

For the three vertices (nodes) of the triangle in Fig. 7.17(a) the potential assumes the following values

$$(7.97) \qquad V_1 = a + bx_1 + cy_1,$$

$$(7.98) \qquad V_2 = a + bx_2 + cy_2,$$

$$(7.99) \qquad V_3 = a + bx_3 + cy_3,$$

or

$$(7.100) \qquad \begin{bmatrix} V_1 \\ V_2 \\ V_3 \end{bmatrix} = \begin{bmatrix} 1 & x_1 & y_1 \\ 1 & x_2 & y_2 \\ 1 & x_3 & y_3 \end{bmatrix} \begin{bmatrix} a \\ b \\ c \end{bmatrix},$$

where (x_1, y_1), (x_2, y_2), and (x_3, y_3) are the coordinates of the vertices; and the determinant of the coefficient matrix in eqn (7.100) may be recognized as being equal to twice the area of the triangle, A. Rearranging eqn (7.100) gives

$$(7.101) \qquad \begin{bmatrix} a \\ b \\ c \end{bmatrix} = \begin{bmatrix} 1 & x_1 & y_1 \\ 1 & x_2 & y_2 \\ 1 & x_3 & y_3 \end{bmatrix}^{-1} \begin{bmatrix} V_1 \\ V_2 \\ V_3 \end{bmatrix},$$

and substitution back to eqn (7.96) yields

$$(7.102) \qquad V = \begin{bmatrix} 1 & x & y \end{bmatrix} \begin{bmatrix} 1 & x_1 & y_1 \\ 1 & x_2 & y_2 \\ 1 & x_3 & y_3 \end{bmatrix}^{-1} \begin{bmatrix} V_1 \\ V_2 \\ V_3 \end{bmatrix}.$$

This last equation may be written as

$$(7.103) \qquad V = \sum_{i=1}^{3} V_i \alpha_i(x, y),$$

where

$$(7.104) \qquad \alpha_1 = \frac{1}{2A} \{ (x_2 y_3 - x_3 y_2) + (y_2 - y_3)x + (x_3 - x_2)y \},$$

$$(7.105) \qquad \alpha_2 = \frac{1}{2A} \{(x_3 y_1 - x_1 y_3) + (y_3 - y_1)x + (x_1 - x_3)y\},$$

$$(7.106) \qquad \alpha_3 = \frac{1}{2A} \{(x_1 y_2 - x_2 y_1) + (y_1 - y_2)x + (x_2 - x_1)y\}.$$

At the vertices,

$$\alpha_1(x_1, y_1) = \frac{1}{2A} \{(x_2 y_3 - x_3 y_2) + (y_2 - y_3)x_1 + (x_3 - x_2)y_1\}$$

$$(7.107) \qquad = \frac{2A}{2A} = 1,$$

$$(7.108) \qquad \alpha_1(x_2, y_2) = \frac{1}{2A} \{(x_2 y_3 - x_3 y_2) + (y_2 - y_3)x_2 + (x_3 - x_2)y_2\} = 0,$$

$$(7.109) \qquad \alpha_1(x_3, y_3) = \frac{1}{2A} \{(x_2 y_3 - x_3 y_2) + (y_2 - y_3)x_3 + (x_3 - x_2)y_3\} = 0$$

and similarly for α_2 and α_3. In general

$$(7.110) \qquad \begin{aligned} \alpha_i(x_j, y_j) &= 0 \quad (i \neq j) \\ &= 1 \quad (i = j); \end{aligned}$$

that is, each function vanishes at all vertices but one, where it assumes the value of unity.

We can now associate energy with each element, and remembering that in the two-dimensional field this energy will be taken per unit length we write

$$(7.111) \qquad W^{(e)} = \tfrac{1}{2} \int_e \varepsilon |\nabla V|^2 \, dS,$$

where the integration is performed over the element area. The potential gradient within the element is found from eqn (7.103) as

$$(7.112) \qquad \nabla V = \sum_{i=1}^{3} V_i \nabla \alpha_i(x, y),$$

so that the element energy becomes

$$(7.113) \qquad W^{(e)} = \tfrac{1}{2} \varepsilon \sum_{i=1}^{3} \sum_{j=1}^{3} V_i \left(\int_e \nabla \alpha_i \cdot \nabla \alpha_j \, dS \right) V_j.$$

Equation (7.113) may be written in the following compact form

$$(7.114) \qquad W^{(e)} = \tfrac{1}{2} \varepsilon [V]^{\mathsf{T}} [N]^{(e)} [V],$$

where $[V]$ is the vector or the vertex values of the potential, the superscript

T denotes transposition, and the 3×3 square element matrix $[N]^{(e)}$ is defined by

$$(7.115) \qquad N_{i,j}^{(e)} = \int_e \nabla \alpha_i \cdot \nabla \alpha_j \, dS.$$

For any given triangle, the matrix $[N]$ is readily evaluated. First, we find the gradients of the α-functions. From eqns (7.104), (7.105), and (7.106)

$$(7.116) \qquad \nabla \alpha_1 = \frac{\partial \alpha_1}{\partial x} \hat{x} + \frac{\partial \alpha_1}{\partial y} \hat{y} = \frac{1}{2A} \{(y_2 - y_3)\hat{x} + (x_3 - x_2)\hat{y}\},$$

$$(7.117) \qquad \nabla \alpha_2 = \frac{\partial \alpha_2}{\partial x} \hat{x} + \frac{\partial \alpha_2}{\partial y} \hat{y} = \frac{1}{2A} \{(y_3 - y_1)\hat{x} + (x_1 - x_3)\hat{y}\},$$

$$(7.118) \qquad \nabla \alpha_3 = \frac{\partial \alpha_3}{\partial x} \hat{x} + \frac{\partial \alpha_3}{\partial y} \hat{y} = \frac{1}{2A} \{(y_1 - y_2)\hat{x} + (x_2 - x_1)\hat{y}\}.$$

Notice that for the first-order approximation, given by eqn (7.96), the gradients of α-functions are constant within an element. Now, the scalar product of two vectors, say a and b, in a Cartesian-coordinate system in two dimensions is given by

$$\begin{aligned}
a \cdot b &= (a_x \hat{x} + a_y \hat{y}) \cdot (b_x \hat{x} + b_y \hat{y}) \\
&= a_x b_x \hat{x} \cdot \hat{x} + a_x b_y \hat{x} \cdot \hat{y} + a_y b_x \hat{y} \cdot \hat{x} + a_y b_y \hat{y} \cdot \hat{y} \\
(7.119) \qquad &= a_x b_x + a_y b_y.
\end{aligned}$$

The scalar products of the gradients of α-functions can therefore be easily found. As these gradients are constant within an element, their scalar products will also be constant. Hence integration over the element area will introduce the triangle area, A, as a constant multiplier. The elements of the matrix $[N]$ can now be found, with a typical expression in the form

$$N_{11}^{(e)} = \frac{1}{4A} \{(y_2 - y_3)^2 + (x_3 - x_2)^2\},$$

$$N_{12}^{(e)} = \frac{1}{4A} \{(y_2 - y_3)(y_3 - y_1) + (x_3 - x_2)(x_1 - x_3)\},$$

$$(7.120) \qquad N_{13}^{(e)} = \ldots .$$

Other entries can be obtained by cyclic permutation of subscripts.

This completes the specification for an arbitrary element in the finite-element mesh. The total energy associated with the entire region will be found as the sum of individual element energies

$$(7.121) \qquad W = \sum W^{(e)}$$

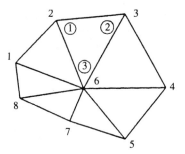

Fig. 7.20 Global and local node numbering.

for all elements. When assembling such elements it can be immediately noticed that some nodes will be shared between more than one element, as shown in Fig. 7.20, and thus the topology of the actual mesh will directly affect the way in which the global matrix is formulated. In other words, the global node numbering must be related to the local numbering, and the global matrix must reflect the way in which individual elements are linked to global nodes. This process is discussed in the next section.

7.10 Discretization and matrix assembly

In order to clarify the procedure, a particular simple example will be used to illustrate the process of assembling the matrix and then solving the system of equations. Consider the region depicted in Fig. 7.21(a). The solution here is trivial and by inspection $V_1 = V_2 = 50$, and the distribution of the field is uniform. However, we shall now apply the general method of solution. The unconstrained nodes are numbered first, and the constrained nodes are numbered last. A particular combination of local and global node numbering schemes is demonstrated in Fig. 7.21(b). This combination has been chosen for convenience, but other combinations are also possible. The data is summarized in Tables 7.1 and 7.2.

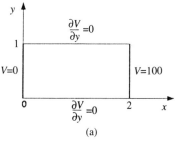

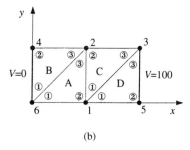

(a)   (b)

Fig. 7.21 Example.

Table 7.1 Coordinates and potentials of nodes

Node number	x-coordinate	y-coordinate	Potential
1	1	0	?
2	1	1	?
3	2	1	100
4	0	1	0
5	2	0	100
6	0	0	0

Table 7.2 Elements and node numbers

Element	Vertex 1	Vertex 2	Vertex 3
A	6	1	2
B	6	4	2
C	1	2	3
D	1	5	3

The element matrices may be determined using eqn (7.120). For example, for element A

$$(7.122) \qquad \begin{cases} x_3 - x_2 = 0, & y_2 - y_3 = -1, \\ x_1 - x_3 = -1, & y_3 - y_1 = 1, \\ x_2 - x_1 = 1, & y_1 - y_2 = 0, \end{cases}$$

and the element area $A = 0.5$. Thus

$$N_{11}^{(A)} = \frac{1}{4A}\{(y_2 - y_3)^2 + (x_3 - x_2)^2\} = \tfrac{1}{2}\{(-1)^2 + (0)^2\} = \tfrac{1}{2}$$

$$N_{12}^{(A)} = \frac{1}{4A}\{(y_2 - y_3)(y_3 - y_1) + (x_3 - x_2)(x_1 - x_3)\}$$

$$= \tfrac{1}{2}\{(-1)(1) + (0)(-1)\} = -\tfrac{1}{2}.$$

$$(7.123) \qquad N_{13}^{(A)} = \dots.$$

Careful inspection of elements B, C, and D shows that, in our example, all the element matrices are the same, that is,

$$(7.124) \qquad [N]^{(A)} = [N]^{(B)} = [N]^{(C)} = [N]^{(D)} = \begin{bmatrix} \tfrac{1}{2} & -\tfrac{1}{2} & 0 \\ -\tfrac{1}{2} & 1 & -\tfrac{1}{2} \\ 0 & -\tfrac{1}{2} & \tfrac{1}{2} \end{bmatrix}.$$

Note that these matrices are equal in our example because of the particular proportions and the node numbering. In general, of course, the element matrices would all be different.

There are six nodes in our mesh, so the global matrix will have dimension of 6×6. Each element matrix will be embedded into the global matrix in a way which depends on the relation between the local vertex numbering and the global node numbering. For example, the elements of the matrix $[N]^{(A)}$ will take the following positions in the global matrix

(7.125)
$$[N]^{(A)} = \begin{bmatrix} N_{66} & N_{61} & N_{62} \\ N_{16} & N_{11} & N_{12} \\ N_{26} & N_{21} & N_{22} \end{bmatrix}$$

The subscripts identify where the contribution should be placed in the global matrix. For example, N_{66} is placed in row six, column six of the matrix. Repeating this process for all four elements, and adding appropriate terms, results in the following global matrix

(7.126)
$$[N] = \begin{bmatrix} 2 & -1 & 0 & 0 & -\frac{1}{2} & -\frac{1}{2} \\ -1 & 2 & -\frac{1}{2} & -\frac{1}{2} & 0 & 0 \\ 0 & -\frac{1}{2} & 1 & 0 & -\frac{1}{2} & 0 \\ 0 & -\frac{1}{2} & 0 & 1 & 0 & -\frac{1}{2} \\ -\frac{1}{2} & 0 & -\frac{1}{2} & 0 & 1 & 0 \\ -\frac{1}{2} & 0 & 0 & -\frac{1}{2} & 0 & 1 \end{bmatrix}.$$

Thus, for any particular topology and node numbering, the global matrix may be found. This process would be too tedious for hand calculations, but it is easily performed by a computer program.

Let us go back to our fundamental formulation. In order to minimize the total-energy expression, eqn (7.121) must be differentiated with respect to a typical value of V_k and then equated to zero. Thus

(7.127)
$$\frac{\partial W}{\partial V_k} = 0,$$

where the index k refers to node numbers in the global numbering scheme. In a boundary-value problem, like the problem in our example, some boundary segments have specified potential values. Thus a subset of the node potentials contained in the vector $[V]$ will assume exactly those prescribed values. We have been careful in our example to number the nodes which are free to vary first, thus leaving all nodes with a prescribed potential to the last. This is not strictly necessary, but it is prudent at this stage. It allows eqn (7.127) to be rewritten with the matrices in partitioned form,

(7.128)
$$\frac{\partial W}{\partial V_k} = \frac{\partial}{\partial [V_f]_k} [[V_f]^T [V_p]^T] \begin{bmatrix} [N_{ff}] & [N_{fp}] \\ [N_{pf}] & [N_{pp}] \end{bmatrix} \begin{bmatrix} [V_f] \\ [V_p] \end{bmatrix} = 0,$$

where the subscript f and p refer to nodes with 'free' and 'prescribed' potentials, respectively. Note that the prescribed potentials cannot vary, and thus differentiation with respect to them is not possible. Hence differentiation with respect to the free potentials results in the following matrix equation

(7.129)
$$[[N_{ff}] [N_{fp}]] \begin{bmatrix} [V_f] \\ [V_p] \end{bmatrix} = 0$$

and leads to a system of algebraic equations of the form $Ax = b$; namely

(7.130)
$$[N_{ff}] [V_f] = -[N_{fp}] [V_p],$$

which has a formal solution

(7.131)
$$[V_f] = -[N_{ff}]^{-1} [N_{fp}] [V_p].$$

In our example

(7.132)
$$[N_{ff}] = \begin{bmatrix} 2 & -1 \\ -1 & 2 \end{bmatrix}$$

and

(7.133)
$$[N_{fp}] = \begin{bmatrix} 0 & 0 & -\frac{1}{2} & -\frac{1}{2} \\ -\frac{1}{2} & -\frac{1}{2} & 0 & 0 \end{bmatrix},$$

so that the final system of equations is given by

(7.134)
$$\begin{bmatrix} 2 & -1 \\ -1 & 2 \end{bmatrix} \begin{bmatrix} V_1 \\ V_2 \end{bmatrix} = -\begin{bmatrix} 0 & 0 & -\frac{1}{2} & -\frac{1}{2} \\ -\frac{1}{2} & -\frac{1}{2} & 0 & 0 \end{bmatrix} \begin{bmatrix} V_3 \\ V_4 \\ V_5 \\ V_6 \end{bmatrix}$$

This is of course a simple system of two equations with two unknowns

(7.135)
$$\begin{cases} 2V_1 - V_2 = 50, \\ -V_1 + 2V_2 = 50 \end{cases}$$

which yields

(7.136)
$$V_1 = V_2 = 50$$

on solution, as expected.

In our example we used a particular simple notation for the system of equations and the nodes had to be numbered in a particular sequence; that is, all potentials free to vary were numbered first, and all potentials with prescribed values were numbered last. In practice, this numbering scheme may not be convenient, and in fact it will not be necessary; it has been used here for purposes of explanation only.

It is also important to note that the finite-element solution is uniquely defined everywhere, not just at the nodes. It is convenient to store the results as a set of nodal potential values, but this is merely a compact representation for the piecewise-planar solution surface which minimizes the system energy.

In conclusion, the finite-element method overcomes the main difficulties of the finite-difference technique, in particular, accurate matching of irregular boundary shapes and higher-order approximation, and offers more flexibility. Only a very basic formulation for Laplace's equation has been presented and illustrated using a simple example. The Galerkin weighted residual technique, mentioned before, is more general than the variational formulation and therefore it is preferred, but in those cases where the energy functional is known the two formulations lead to exactly the same numerical model, that is, an identical set of algebraic equations.

7.11 Solving the system equations

Both the numerical methods introduced in this chapter, the finite differences and finite elements, transform the appropriate partial differential equation to a discretized set of algebraic equations $Ax = b$. Boundary conditions and interface conditions between different materials will also be embedded in this equation. Many approaches exist for obtaining the solutions of such equations. They will often be based on a direct method, such as Gaussian elimination, or on an iterative approach, for example, *successive overrelaxation* (SOR) or *conjugate gradients* (CGs). The choice of method is important from the practical point of view, as the time and storage requirements of different schemes may be significantly different. We shall not dwell on this topic as the various techniques for solving systems of simultaneous equations are well documented in books on numerical methods and there is an extensive literature on the subject. We will, however, briefly discuss one important property of the equations obtained from finite-difference or finite-element methods; namely that these equations tend to be very sparse.

Let us use a simple example to illustrate the principle. Consider the regular finite-element mesh shown in Fig. 7.22. Assume that nodes 1 to 5 (Face 1) and 21 to 25 (Face 2) have prescribed values of the potential and let the position of the diagonals be arbitrary (sloping upwards or

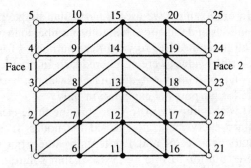

Fig. 7.22 A regular mesh.

downwards). Finally, let the coordinates of the nodes be free to assume any values, so that the mesh is regular in a topological (structural) sense, but the actual object could have a more complicated shape so it does not need to be a rectangle.

The global matrix may be partitioned in the following way

(7.137)
$$[N] = \begin{bmatrix} [N_{pp}] & [N_{pf}] & [N_{pp}] \\ [N_{fp}] & [N_{ff}] & [N_{fp}] \\ [N_{pp}] & [N_{pf}] & [N_{pp}] \end{bmatrix}$$

The $[N_{ff}]$ submatrix is of particular interest. The nonzero entries to this submatrix are shown in Fig. 7.23. This matrix is general for the mesh of quadrilaterals in Fig. 7.22. It shows all the possible links between 'free' nodes. For example, node 14 has possible links with nodes 8, 9, 10, 13, 15, 18, 19, and 20, which can be clearly seen on both figures 7.22 and 7.23. However, once the positions of all diagonals have been chosen, more of the existing entries in the matrix will become zero. For the choice of diagonals in Fig. 7.22, node 14

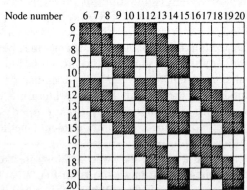

Fig. 7.23 Nonzero entries in the system matrix for the example in Fig. 7.20.

has no link with nodes 10 and 20, so that these two entries in the matrix are zeros. Let us keep the scheme more general, as depicted by Fig. 7.23. The nonzero elements may be numbered. Hence we have

$$(7.138) \quad \begin{bmatrix}
1 \\
16 & 2 \\
\cdot & 17 & 3 \\
\cdot & \cdot & 18 & 4 \\
\cdot & \cdot & \cdot & 19 & 5 \\
36 & 28 & \cdot & \cdot & \cdot & 6 \\
46 & 37 & 29 & \cdot & \cdot & 20 & 7 \\
\cdot & 47 & 38 & 30 & \cdot & \cdot & 21 & 8 \\
\cdot & \cdot & 48 & 39 & 31 & \cdot & \cdot & 22 & 9 \\
\cdot & \cdot & \cdot & 49 & 40 & \cdot & \cdot & \cdot & 23 & 10 \\
\cdot & \cdot & \cdot & \cdot & \cdot & 41 & 32 & \cdot & \cdot & \cdot & 11 \\
\cdot & \cdot & \cdot & \cdot & \cdot & 50 & 42 & 33 & \cdot & \cdot & 24 & 12 \\
\cdot & \cdot & \cdot & \cdot & \cdot & \cdot & 51 & 43 & 34 & \cdot & \cdot & 25 & 13 \\
\cdot & \cdot & \cdot & \cdot & \cdot & \cdot & \cdot & 52 & 44 & 35 & \cdot & \cdot & 26 & 14 \\
\cdot & \cdot & \cdot & \cdot & \cdot & \cdot & \cdot & \cdot & 53 & 45 & \cdot & \cdot & \cdot & 27 & 15
\end{bmatrix}$$

We note that the matrix is symmetrical about the main diagonal; it is sufficient therefore to store the lower (or upper) triangular submatrix. Furthermore, it is not necessary to store the entries which are zero. A simple scheme could be devised to store the nonzero elements in two smaller arrays, as illustrated below. This scheme is not necessarily the most efficient, but is used here to illustrate the principle.

$$(7.139) \quad \begin{bmatrix}
1 & 2 & 3 & 4 & 5 & 6 & 7 & 8 & 9 & 10 & 11 & 12 & 13 & 14 & 15 \\
16 & 17 & 18 & 19 & \cdot & 20 & 21 & 22 & 23 & \cdot & 24 & 25 & 26 & 27 & \cdot
\end{bmatrix} \quad \text{and}$$

$$\begin{bmatrix}
\cdot & 36 & 28 \\
46 & 37 & 29 \\
47 & 38 & 30 \\
48 & 39 & 31 \\
49 & 40 & \cdot \\
\cdot & 41 & 32 \\
50 & 42 & 33 \\
51 & 43 & 34 \\
52 & 44 & 35 \\
53 & 45 & \cdot
\end{bmatrix}$$

Note that the numbers indicate the position, not the value, of appropriate coefficients. These coefficients will have already been calculated using equations derived in the previous section. Moreover, the initial matrix, $[N_{\mathrm{ff}}]$, had $15 \times 15 = 225$ elements. Only 53 of those actually have to be remembered, and in our example we have stored them in two arrays with a total size of $15 \times 2 + 10 \times 3 = 60$.

Most practical programs based on finite differences or finite elements exploit matrix sparsity in order to keep both computing time and memory requirements within reasonable limits. There are many clever techniques and programming 'tricks' employed to achieve maximum savings. Such schemes will probably be much more efficient than the one presented in this section. Typically, the nodes are assumed to be randomly numbered initially, and some systematic renumbering scheme is then applied to produce an improved sparsity pattern.

7.12 Boundary elements

The finite-difference and finite-element methods both depend upon the division of space into a large number of small elements. An alternative formulation, based on integral methods, relies on discretization of active material regions only, or, if the materials are magnetically linear (have constant permeabilities), on discretization of material boundaries only. Two methods in particular are gaining increasing popularity; the first is known as the boundary element method (BEM), it is based on a direct application of Green's second theorem (see Appendix 2, Section A2.5). Another method, known as the boundary integral method (BIM) has some advantages, but it is limited to pure ferromagnetic or pure dielectric problems. The mathematical formulation of these methods is beyond the scope of this book and we shall restrict the discussion to some general comments.

There are several advantages of integral formulations compared with the differential approach using finite-difference or finite-element methods. First, only active regions need to be discretized; this offers enormous advantages in three-dimensional problems, as well as in problems where a high proportion of space is filled with air or other passive materials. Secondly, the far-field condition (in the so-called open-boundary problems) is automatically included in the formulation – unlike in the finite-element method, where very special techniques need to be employed to take account of such conditions. Finally, the fields obtained from the solution are usually very smooth compared with other methods.

The main disadvantage of the integral formulation is that one has to work with a full matrix instead of a sparse matrix and thus the computational costs are high. However, much of the computation involves parallel operations and the use of parallel processing may make the method more attractive. There is increasing interest in the application of the integral methods to electromagnetics. There is also a growing number of hybrid formulations mixing finite elements and boundary elements. This is one of the areas of active research.

7.13 Tubes and slices and finite elements

We have seen how the finite-element method divides regions into small elements, which in two-dimensional problems are generally triangles. The method uses an energy functional which is minimized and leads to a field approximation. This approach is substantially different from the method of tubes and slices, because the elements cannot be treated individually but must always be treated as a complete set. Accordingly, the finite-element method of calculation is based on the solution of a set of simultaneous equations, which is usually a cumbersome process. At the beginning of this process, the information is confined to the boundary elements and the process has to transfer information into the interior of the region. Hence, the finite-element calculation could be greatly accelerated if approximate information about the inner field was available at the beginning. This suggestion leads to a useful combination of tubes and slices with finite elements. First, an approximate field distribution is calculated by the

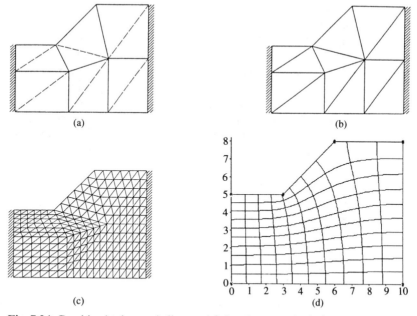

Fig. 7.24 Combined tubes-and-slices and finite-element calculations: (a) construction lines for tubes and slices, (b) a basic finite-element mesh, (c) a refined finite-element mesh, and (d) a field plot obtained from the finite-element solution.

tube/slice method, and then the accuracy is improved by means of a finite-element procedure, which is now significantly simplified. Such a combination of these two powerful techniques is offered in the TAS program accompanying this book. It should be noted that the finite-element method has also been adapted in the TAS program to produce upper and lower bounds to the unkown exact values.

Figure 7.24 demonstrates how this combination of methods works using the example from Section 1.5. The tube/slice distribution in Fig. 7.24(a) is used as a base mesh for the finite-element method, Fig. 7.24(b); this mesh is subsequently refined, Fig. 7.24(c), and values of the solution potential at the nodes are calculated using the tubes-and-slices approximation. This set of node potentials is then refined iteratively using a specially adapted finite-element formulation. Only a small number of iterations is required as the tubes-and-slices solution by itself is quite accurate. The resultant field distribution may then be plotted, Fig. 7.24(d).

7.14 The CAD environment

The role of electromagnetic-field computation in engineering design is an area of very active research, at both academic institutions and specialist software developers, and it is attracting increasing interest from existing and potential users of computational tools for a wide range of technical applications. Many branches of industry have already incorporated computer-aided design (CAD) in electromagnetics into their everyday design practice. The reader is encouraged to consult specialist books on the subject. In this section we shall review briefly some of the fundamental issues and discuss some practical aspects of the CAD environment.

In order to incorporate field computation into the design process, the designer first formulates a mathematical model of his physical problem. This is an absolutely crucial stage which may decide between the success or failure of the whole process. Clearly, if the model is not adequate, no matter how accurately the calculations are conducted, the results will not be of any use. This underlines the importance of human input to the design. It also encourages the designer to conduct experiments and to seek alternative ways of solving the same problem, so that comparisons can be made. Once the model has been formulated, the CAD system will facilitate the finding of the solution. A typical CAD system is illustrated in Fig. 7.25.

There are three stages in the CAD process:

(1) *preprocessor* for defining the geometry, for the input of material data and the definition of the sources and the boundary conditions;
(2) *solver* for performing computation, for example, solution of the system of algebraic equations in the finite-element formulation, or calculation

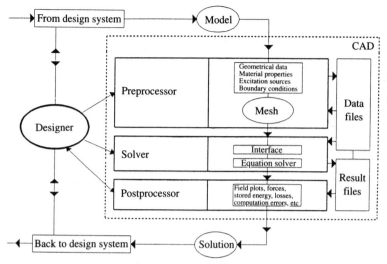

Fig. 7.25 A typical CAD system for electromagnetics.

of the series/parallel connection of components in the tubes and slices formulation; and

(3) *postprocessor* for graphical display of the field distribution; and for calculation of the field components and integrating along lines, over areas or inside volumes, to find integral parameters such as forces, stored energy, impedances, etc.

It is helpful and convenient if both preprocessing and postprocessing are interactive, taking full advantage of the graphics capabilities of modern computers. The solution, on the other hand, is very often 'hidden' from the user and performed using batch processing or sent, via a network, to a mainframe computer. In the TAS program, on the other hand, due to the efficient formulation of the method, the solver is effectively amalgamated with the rest of the graphics interface, because the solution times are extremely short.

Software based on the finite-element formulation offers various levels of automation of mesh generation. In an ideal system discretization would in fact form part of the solution stage, rather than preprocessing, as shown in Fig. 7.26. This is known as *adaptive meshing*, and some existing programs already offer this facility, although it is perhaps not yet fully integrated into the design process. Adaptive meshing is a rather difficult concept for general application, and it is an area of active research.

Most finite-element programs use three types of mesh generation: deterministic, semi-automatic and interactive. The choice often depends upon types of geometric 'primitives' (that is, basic geometrical shapes)

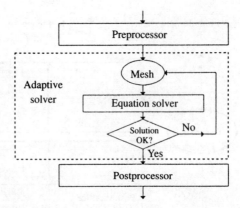

Fig. 7.26 Adaptive meshing.

which are used to construct the finite-element model. At the same time, the designer may wish to have full control over the element distribution, for example, to take proper account of the skin effect. A 'manual' input may be easily implemented for a quadrilateral, where the definition of boundary subdivisions results in a predetermined internal meshing. Some simple examples in two dimensions are illustrated in Fig. 7.27. Clearly, there are more possibilities, and the principle applies not only to rectangles but to any quadrilateral.

Semi-automatic methods apply to a polygon of any shape, but the subdivision of edges is usually predetermined. The problem is to find the optimum distribution of internal points, and then the best way of connecting these points. The intuitive criterion is that all triangles should be as near to equilateral as possible, to avoid large obtuse angles and produce a mesh which is well proportioned. Several algorithms have been developed and they are well covered in specialist books on finite elements. Probably the most popular method is the so-called *Delaunay triangulation*. An illustration of a 'bad' and a 'good' mesh for a given set of points is given in Fig. 7.28.

Interactive methods allow particular 'nodes' or 'construction lines' of the mesh to be repositioned directly on the screen of the terminal using a

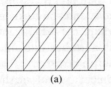 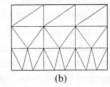 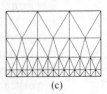

(a)　　　　　　　　(b)　　　　　　　　(c)

Fig. 7.27 Simple meshes for quadrilaterals: (a) regular, (b) graded in one direction, (c) graded and nonuniform.

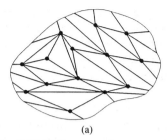
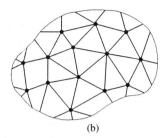

(a)   (b)

Fig. 7.28 Different meshes constructed on the same set of points, (a) A 'bad' mesh, and (b) A 'good' mesh.

graphics interface, for example, a mouse. This method is usually used to complement other methods, to apply finishing touches to the manually or automatically generated mesh.

In addition, most commercially available programs offer useful extra features for preprocessing, such as copying, replication, rotation, mirror images, loop inputs, as well as libraries of shapes, material properties, magnetization curves, etc. Moreover, the preprocessor usually performs several checks on the generated mesh to assure continuity of element subdivision, completeness of the material data, matching of interface boundaries, sufficiency of the applied boundary conditions, and other factors. These checks are often highly automated.

The solver is, in a sense, the most important part of the CAD process, although interestingly it is the part which is the least visible to the designer and which does not usually involve the designer. The process is initiated by the user (by executing a batch file, choosing an option from the menu, etc) and at the end of the solution the user is put into the postprocessing mode. The solution itself may be a simple process (for example, in tubes and slices) or it may entail the solution of a set of simultaneous equations. We have already discussed these matters in some detail in previous sections.

Versatile postprocessing is crucial from the practical point of view to facilitate reading and understanding of the results. At this stage the solution is probably already known in the form of hundreds or thousands of values of solution potentials at discrete points of the system. This information has to be translated into meaningful engineering quantities. This will involve differentiation, integration (line, area, or volume), contour plotting, and other special numerical or graphical techniques. Some postprocessors incorporate smoothing techniques to refine the quality of the output information. Designers need to be able to 'see' the solution – this is typically provided by field plots such as displays of tubes and slices, coloured displays as area zones of field levels, vector displays showing the magnitude and direction of field vectors graphically as arrows at selected points, and line displays showing the variation of some field quantity along a selected path. But the designer also needs global (circuit) parameters, such as the resistance, inductance, or capacitance of some circuit element. In TAS this information is provided automatically. In most programs, stored energy and power loss have to be calculated, and this involves volume or area integration. Surface or line integration may also be necessary when calculating forces.

Very often the results from an electromagnetic computation will be used in other calculations. For example, forces from electromagnetic analysis could be used in mechanical-stress analysis; or distributed power loss would be a source of a particular thermal field. The postprocessor must be capable of transferring such useful information further into the design process, so that this information can then be read by other programs.

Finally, two particular aspects deserve special mention. First, methods for *error estimation* become an increasingly important and active area of research. Error estimators allow the user to predict which part of the solution is subject to the biggest errors so that improvements can be made by appropriate mesh modifications. Methods based on dual energy bounds appear to be particularly promising. Error estimation is an inherent component of the adaptive-meshing procedure. And secondly, optimization techniques are gradually being introduced to CAD in electromagnetics. Some degree of success has already been achieved, but these are still very early days.

All the above requirements are very demanding, with often conflicting objectives in terms of accuracy, speed and convenience of calculation, user-friendliness, levels of automation and user control, support for different hardware and software systems, and data transfer. It is not surprising, therefore, that professional electromagnetic software requires many years of research and development and existing programs are very complex. However, in the hands of an expert user they become extremely powerful tools, and CAD in electromagnetics has already been established as an everyday design practice in many branches of industry.

Exercises

7.1 A dielectric strip of uniform thickness and of constant width, d, (see Fig. 7.29) has a potential $V_0 \sin(\pi x/d)$ applied at the end $y=0$. The strip extends from $y=0$ to $y \to \infty$. The sides are kept at ground potential, and the thickness of the strip is such that it does not affect the field distribution; that is, the field may be considered as two-dimensional. Show that the potential inside the strip is given by

$$V = V_0 \exp\left(-\frac{\pi y}{d}\right) \sin\left(\frac{\pi x}{d}\right)$$

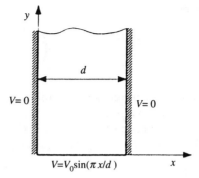

Fig. 7.29 A dielectric strip.

and the electric field strength is described by

$$E = \frac{\pi V_0}{d} \exp\left(-\frac{\pi y}{d}\right)\left\{-\cos\left(\frac{\pi x}{d}\right)\hat{x} + \sin\left(\frac{\pi x}{d}\right)\hat{y}\right\}.$$

7.2 Consider the rectangular parallelepiped in Fig. 7.30 with dimensions a, b, and c. A sinusoidal potential distribution

$$V_0 \sin\left(\frac{\pi x}{a}\right)\sin\left(\frac{\pi y}{b}\right)$$

is applied to the top of the parallelepiped, and the other faces are kept at zero potential. Show that the potential distribution within the brick is given by

$$V = V_0 \frac{\sinh(\lambda z)}{\sinh(\lambda c)} \sin\left(\frac{\pi x}{a}\right)\sin\left(\frac{\pi y}{b}\right),$$

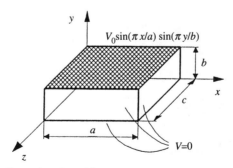

Fig. 7.30 A three-dimensional problem.

where

$$\lambda = \pi\left(\frac{1}{a^2} + \frac{1}{b^2}\right)^{1/2}.$$

7.3 A solid magnetic cylinder of radius R is placed in a uniform transverse magnetic field B_0 (see Fig. 7.31). The magnetic material is linear and its relative permeability is μ_r. Show that the field inside the cylinder is uniform (see Fig. 7.32) and that the flux density inside the cylinder is given by

$$B_{in} = B_0 \frac{2\mu_r}{\mu_r + 1}.$$

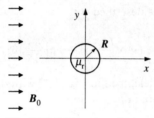

Fig. 7.31 Magnetic cylinder.

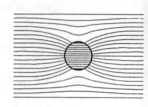

Fig. 7.32 Flux plot.

7.4 Discuss the use of electric and magnetic images for calculating fields due to charges or currents in the vicinity of conducting or magnetic bodies. Suggest a system of multiple images for representing the magnetic field of a line current placed in a slot in iron of permeability μ_r, as shown in Fig. 7.33.

(*Answer* a possible system is shown in Fig. 7.34, where $M = (\mu_r - 1)/(\mu_r + 1)$.)

Fig. 7.33 The line current in a slot in iron.

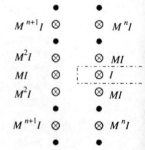

Fig. 7.34 Multiple images.

7.5 Figure 7.35 shows a current filament of strength I in the right angle between two large iron surfaces. Estimate the force on the current assuming that the iron has a very high permeability.
Answer

$$\left(F_x = -\frac{\mu_0 I^2 (2a^2 + b^2)}{4\pi a(a^2 + b^2)}, \qquad F_y = -\frac{\mu_0 I^2 (a + 2b^2)}{4\pi b(a^2 + b^2)}. \right)$$

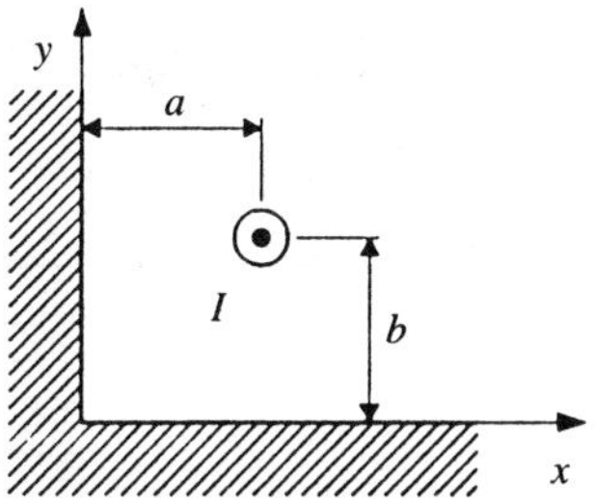

Fig. 7.35 The current near a magnetic block.

7.6 Explain why it is possible to determine an electrostatic field in a nonconducting dielectric by measurements on a conducting sheet or electrolytic tank. Discuss how other fields can be represented by means of a conducting analogue.

7.7 Explain how the distribution of flux in a transformer stamping can be investigated by means of a network analogue. How would you represent the condition that: (a) the surface magnetic field strength is given, or (b) the total flux is given?
(*Answer RC* network; (a) voltage at boundary; (b) total current.)

7.8 Discuss the method of *curvilinear squares* and compare it with the method of *tubes and slices*.

7.9 Show that Laplace's equation in three dimensions in an x, y, z coordinate system may be represented using a seven-point finite-difference scheme

$$\phi_{i,j,k} = \tfrac{1}{6}(\phi_{i-1,j,k} + \phi_{i+1,j,k} + \phi_{i,j-1,k} + \phi_{i,j+1,k} + \phi_{i,j,k-1} + \phi_{i,j,k+1}),$$

where it has been assumed that distances between adjacent nodes are the same; that is, $\Delta x = \Delta y = \Delta z$.

7.10 Find the unknown potentials of nodes 1 to 5 for the example discussed in Section 7.8 (see Fig. 7.15) by solving the system of equations described by eqn (7.81).

(*Answer* the calculated potentials are shown in Fig. 7.36. The values in parentheses represent the accurate solution, which may be obtained using a fine mesh or a different method – refer to Exercises 7.13 and 7.15.)

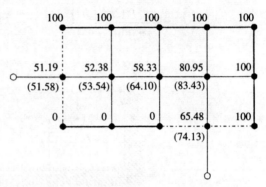

Fig. 7.36 Calculated node potentials.

7.11 Use the five-point scheme of eqn (7.79) to derive a simple iterative scheme for calculating unknown potentials. Hence demonstrate how the solution converges on the values obtained in Exercise 7.10.
(*Answer* a short computer program will be helpful. A possible program written in C is suggested below. The values computed after one, two, and nine iterations are shown in Fig. 7.37.)

```
⋮
v1=0.0; v2=0.0; v3=0.0, v4=0.0; v5=0.0; iter=0;
do{
   v1=0.25*(100+2*v2);
   v2=0.25*(100+v1+v3);
   v3=0.25*(100+v2+v4);
   v4=0.25*(200+v3+v5);
   v5=0.25*(100+2*v4);
   iter+ =1;
   printf("%3d %6.2f %6.2f %6.2f %6.2f %6.2f\n",iter,v1,v2,v3,v4,v5);
} while (iter<20);
⋮
```

7.12 It has been found that the rate of convergence of the iterative scheme illustrated by Exercise 7.11 could be improved by modifying the values calculated in each iteration. The 'successive overrelaxation' scheme uses a 'relaxation factor', α, in the following way

$$V_i^{\text{new}} \Leftarrow V_i^{\text{old}} + \alpha(V_i^{\text{new}} - V_i^{\text{old}})$$

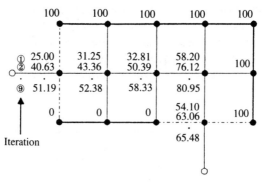

Fig. 7.37 Results from the simple iterative scheme.

for each node, where *old* refers to values calculated in the last iteration, and *new* is the current iteration. The relaxation factor is restricted to $1 \leq \alpha < 2$. Modify the calculations of Exercise 7.11 by using this new scheme with $\alpha = 1.15$. Try other values of α to see their effect on the convergence.

(*Answer* a modified section of the computer program is shown below. Potentials v2 to v5 will be modified in a similar way. Note that the use of the variable 'vold' is not essential but makes the program easier to read. The values computed after iterations one, two, and seven are shown in Fig. 7.38.)

```
⋮
vold = v1;
v1 = 0.25*(100 + 2*v2);
v1 = vold + alpha*(v1 − vold);
⋮
```

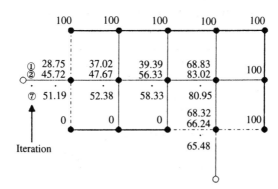

Fig. 7.38 Results from the successive over-relaxation scheme.

7.13 Write a more general computer program, for the same geometry, using the ideas from Exercise 7.11 and 7.12 for finite-difference calculations based on a finer mesh of fixed topology. A possible numbering sequence is shown in Fig. 7.39, where n is the number of node layers in the vertical direction. Experiment with various values of n to see how fine the mesh must be to achieve the accuracy suggested in Fig. 7.36.

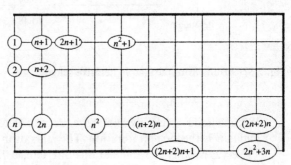

Fig. 7.39 Node numbering for a fine mesh.

7.14 Use the method of *tubes and slices* of the TAS program to calculate the capacitance of the system studied in Exercises 7.10 to 7.13. Experiment with the number and distribution of tube and slice lines to see their effect on the accuracy of the calculated dual bounds.

7.15 Use the *finite-element method* of the TAS program and find the potential and field distributions for the capacitive problem of the previous exercises. Experiment with the number of elements used for the solution and its effect on the accuracy of computation. In the postprocessor, use the calculation of capacitance as well as the display of field values at points in order to assess the accuracy.
(*Answer* the finite-element solution is given in Fig. 7.40.)

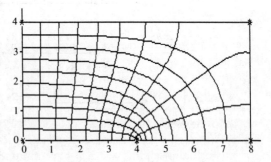

Fig. 7.40 A finite-element solution for the problem of Fig. 7.15.

Engineering applications   8

8.1 Introduction

The range of application of electromagnetism is enormous. Without
electromagnetic devices and systems it would be impossible to maintain the
earth's human population, since both industry and agriculture depend on
the application of electromagnetic-energy processes. Although the use of
electromagnetics does not generally require a knowledge of the principles
described in the previous chapters of this book, such knowledge is required
by many of the engineers engaged in design and development work. There
are of course many different aspects and specializations in such work. A big
field is that of telecommunications, involving antennas, satellite systems,
radar, radio astronomy, mobile communication systems, and microwave
links. In electric-power applications one can mention the design of large
rotating machines and transformers, high-voltage switches and insulation
systems, and cables and transmission systems. Electric motors are used in
immense numbers and in a wide variety of situations including domestic
equipment, ancillary equipment in motor cars, drives and control
mechanisms in factories, railways, and mobile machinery. Electrostatics is
important in powder and paint deposition, the protection of electronic
circuits and the avoidance of explosive hazards in flour mills, and the
fuelling of aircraft. Eddy currents are used in induction furnaces. A growing
field of electromagnetic devices is in medicine, including noninvasive
diagnostic tools such as magnetic-resonance-imaging magnets. The list is
practically endless.

All these applications have a specialist literature and it is impossible to
include a sufficient number of examples to be regarded as an adequate
survey. In any case, the authors have only limited experience, and
university courses cannot do more than deal with a selection of the most
important topics. Nevertheless, we think that the inclusion of a final
chapter dealing with applications will be helpful to the reader by showing
how the principles and methods of electromagnetism can be used in
practice. Although a good case can be made for the teaching of the subject
as an intellectual discipline, we also have in mind its usefulness; and this
chapter is intended to stress such usefulness.

8.2 An inductive sensor: a case study

In this section we shall demonstrate how the methods of electromagnetics can be applied to analyse the operation of a practical device. We shall begin by discussing the physical behaviour of the device. We shall then formulate an appropriate computational model and discuss various assumptions. Finally, we shall look into the possibility of simplifying our computation, and we shall assess the consequences of the approximations introduced. The whole process should be understood as forming a part of a design procedure with the objective, for example, of optimizing the performance of the device. The final stage in that process would be the building and testing of the device in order to validate the modelling techniques used and the simplifications adopted.

As an illustration, we shall use a popular inductive sensing device of the type shown in Fig. 8.1. Such devices operate on a principle of detecting changes in the *reflected impedance* of the coil as a result of a field disturbance due to the proximity of a magnetic/conducting target. The electronic circuit is not shown, but its simple task is to detect or measure the changes in the *resistance* and *inductance* of the coil (or a combination of the two, for example, an impedance). Such changes may be used to calibrate the device. Applications include proximity detectors, inductive sensors, coin-recognition systems, and similar devices.

Let us briefly discuss the electromagnetics of the device. The magnetic circuit provided by the ferrite core is not closed, and thus the field level and distribution will depend mainly on the *reluctance* of the air space above the core in the direction of the target. The relative permeability of an unsaturated ferrite core is typically around 1000. If there is no target, the magnetic field produced by the current flowing in the coil will produce the field distribution illustrated in Fig. 8.2. The flux lines are confined to the magnetic-core region for part of their path, but they have to pass through the air space between the core 'rims' to complete the magnetic circuit. The

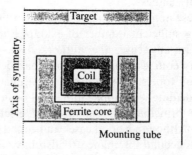

Fig. 8.1 An inductive sensor.

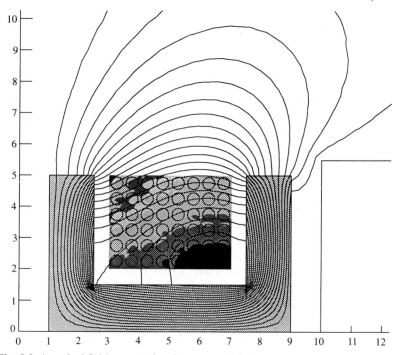

Fig. 8.2 A typical field pattern for the sensor without a target.

ampere-turns of the coil provide the source of the field (m.m.f.) and the equivalent circuit of the system will be similar to that discussed in Section 3.6. However, the shape of the 'air gap' is now much more complicated; and we think that the notion of reluctance is helpful in aiding understanding of the process, but it is not sufficient to perform accurate calculations. We shall resort to numerical analysis and finite-element modelling to analyse the system. (The finite-element calculations reported in this chapter were conducted using the package OPERA-2D, which was developed by Vector Fields Ltd, Oxford, UK.)

The magnetic flux linkage with the current of the coil gives rise to the *inductance* of the coil. However, as seen from Fig. 8.2, this linkage is now somewhat complicated as some of the flux actually passes through the coil and thus different coil turns will have a slightly different linkage. Computationally, it is more convenient to calculate the inductance via the field energy. Nevertheless, the concept of flux linkage will continue to be helpful in our discussion. In the example, we have assumed that the coil has 48 turns.

If a conducting and/or magnetic target is brought into the proximity of the sensor, as shown in Fig. 8.1, the magnetic-field distribution will be affected, and there are two mechanisms which will contribute to the changes.

1. *The magnetic effect* If the target is made from magnetic but non-conducting material of high permeability, the total reluctance of the system will be reduced thus increasing the total flux generated by the same ampere-turns. The inductance of the coil will be increased.
2. *The electric effect* In a nonmagnetic but conducting target, *eddy currents* will be induced repelling the flux and thus forcing it to go through a region of smaller cross-sectional area. This will cause a reduction in the value of the total flux and thus a reduction of the inductance. Both effects are illustrated later in this section (Figs 8.7 and 8.8).

Some targets are likely to fall into one of the above categories; copper or aluminium materials, for example, will exhibit only electric effects. An *iron target*, on the other hand, will produce both effects at the same time. In an unsaturated iron, a dominating magnetic effect will be hindered by the opposing influence of eddy currents. As the electric effect becomes stronger with increased frequency, the magnetic flux will be confined to a very thin surface 'skin', with a considerably increased surface flux density. A measure of this effect is given by the skin-depth parameter, $\Delta = (\omega\mu\sigma/2)^{-1/2}$ (refer to Section 5.7). There will be little flux in the interior of the iron beyond the skin. The high flux density in the skin will lead to local magnetic saturation which is likely to reduce the 'effective' permeability of the iron, although the increased value of Δ will act against this effect somewhat.

From the above physical insight, it emerges that a variation of the inductance with the target distance is to be expected, subject to different material properties of the target combined with the frequency of excitation. Magnetic but nonconducting materials should give an increase in the coil inductance as the target is brought closer to the coil. Purely conducting (nonmagnetic) targets will cause the inductance to fall. Iron targets may produce a variety of responses depending on the frequency. For higher frequencies, due to the skin effect and local saturation, the two opposing effects (magnetic and electric) may even cancel out, giving virtually no change of inductance with target range, although there will be a change in resistance.

The effective resistance of the coil consists of the resistance of the coil itself at the given frequency (with the possibility of a *skin effect* in the conductor, a *proximity effect* in the coil and an *eddy-current effect* in other conducting parts of the system – all these effects are explained later in this section) plus a term representing power loss in the conducting target, which is always positive. This power has to be provided by the current in the coil, and thus the supplying system will detect an effective increase in the coil resistance. In most cases the overall effect (that is, including the changes to the other losses) of bringing in the target will be an increase in the coil resistance.

Before we proceed to actual calculations, we need to formulate an appropriate computational model, and there are some very interesting properties which are worth investigating. We must not forget that the usefulness of the results depends crucially on the relevance of the model, but at the same time there is little point in trying to model accurately those aspects or details of the device which have a negligible effect on the important parameters. This is where the knowledge and experience of the designer become irreplaceable and where the failure or success of the design process is determined.

The first problem which we will address is relevant to most magnetic devices containing coils. Coils are wound with a number of turns of a conductor and the question arises whether it is necessary that all the individual conductors should be modelled. For our example (where we have assumed 48 turns) this is just possible, as shown in Fig. 8.3, but it does produce a jungle of small elements in the coil area and it is rather impractical.

Most of the cross section of the coil is filled with conductors, and they all

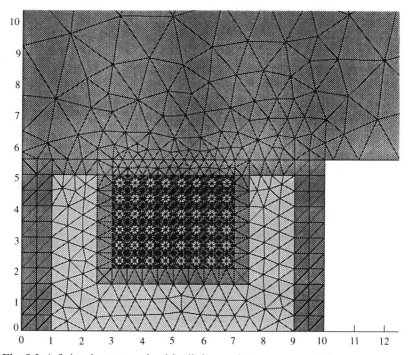

Fig. 8.3 A finite-element mesh with all the conductors represented.

carry the same current i, so that the ampere-turns of the coil are ni, where n is the number of turns. For the purpose of the computation of a magnetic field we can ignore the presence of individual conductors and replace them with one big conductor with a uniform current-density distribution, giving the total current as equal to the ampere-turns of the coil. This leads to a dramatic reduction in the number of elements required for finite-element modelling. Figure 8.4 demonstrates the new mesh, and Fig. 8.5 shows the resulting field distribution. Comparison with Fig. 8.2 shows that the two field patterns are virtually identical, and thus the simplification is justified. This is one of the popular and generally accepted 'tricks' used to simplify calculations. However, this simplification has some impact on the calculation of the coil *proximity loss*; we shall come back to this problem later in this section.

The skin depth of an iron target will be particularly small under typical operating frequencies and it needs to be modelled using a very fine graded mesh (see Fig. 8.4). Combined with the need for taking account of the nonlinear magnetic characteristics of iron, this increases computing times considerably. One possible simplification might be to use linearized magnetic characteristics of iron so that saturation is either ignored or

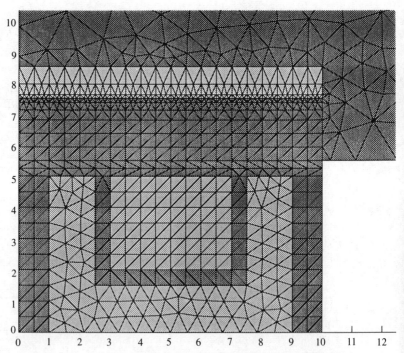

Fig. 8.4 A finite-element mesh with a meshed target and a simplified coil.

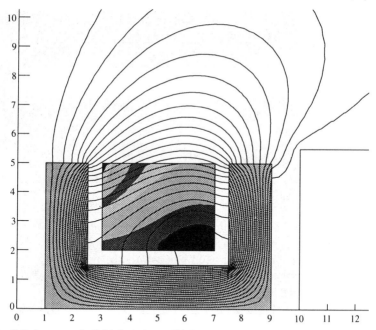

Fig. 8.5 A magnetic field due to one 'big' turn.

represented by using an 'effective' permeability. An estimate could be made by first solving a static problem and then assuming that the same total flux must be pushed through the skin. This new permeability can then be used for subsequent eddy-current calculations. Numerical modelling of thin skin depths is generally a difficult problem.

The presence of a nonmagnetic conducting target at higher frequencies having a small skin depth can be modelled in an approximate way by a 'cavity', with the boundary condition $A = 0$ applied, where A is the magnitude of the vector potential. Such a condition prevents the magnetic flux from entering the 'cavity' in the way real copper or brass prevents the entry of flux at high frequency. A much simpler mesh is then sufficient. Figure 8.6 illustrates a typical example. (Comparison between the field plots of Figs 8.8 and 8.9 shows that the 'cavity' model is perfectly adequate.) The cavity model has also been used to represent the presence of the conducting mounting tube; that is, the condition $A = 0$ has been applied to all surfaces of this tube.

The resulting finite-element models are well suited for interactive CAD purposes. Examples of field plots for different types of targets are shown in Figs 8.7 to 8.10. The calculated field distributions confirm the theoretical predictions. The main advantage of having such field plots is that they

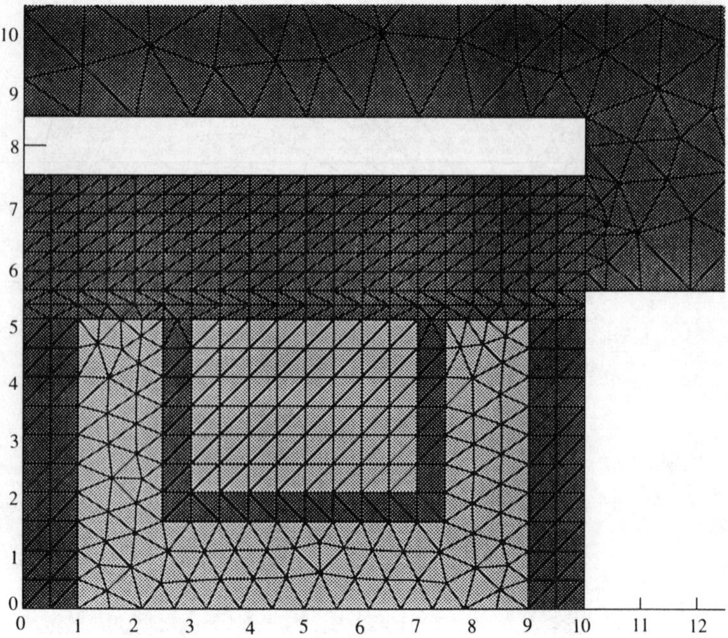

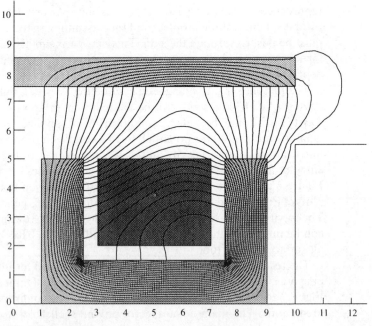

Fig. 8.6 A 'cavity' model.

Fig. 8.7 A field distribution for a magnetic target.

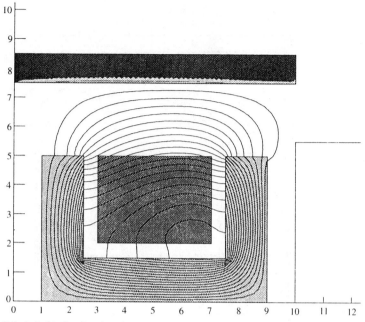

Fig. 8.8 A field distribution for a conducting target.

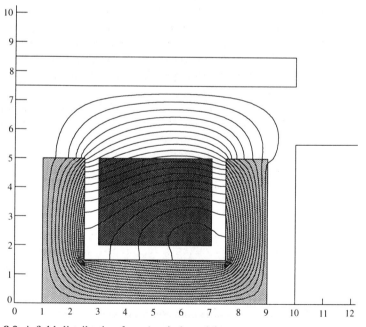

Fig. 8.9 A field distribution for a 'cavity' model.

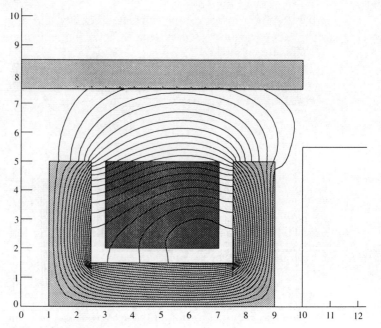

Fig. 8.10 A field distribution for an iron target.

provide insight and visual interpretation of the behaviour of the field. Such insight is rarely available to the designer without computer-visualization techniques. In the following calculations a typical sensor geometry has been assumed and the dimensions have been chosen to emphasize various effects. Similar calculations conducted for commercial sensors have shown good agreement with experimental data, thus validating the techniques and simplifications adopted.

The inductance of the coil for different target positions can be found from the finite-element solution through integration of the stored (or supplied) energy, W. An appropriate volume-integration facility is usually provided within the commercial software. The stored energy is expressed in terms of the inductance by $W = \frac{1}{2}Li^2$, from which L may be easily found.

The 'effective' resistance of the coil may be written as

(8.1) $$R = R_{\text{d.c.}} + R_{\text{skin}} + R_{\text{prox}} + R_{\text{other}} + R_{\text{target}},$$

where $R_{\text{d.c.}}$ is the d.c. (direct current) resistance (that is, the resistance measured using d.c. current), R_{skin} represents the *skin effect* in individual conductors, R_{prox} is due to the *proximity effect* (currents induced in a conductor by the field of other conductors), R_{other} includes power loss in all other conducting parts of the sensor (for example, in the mounting tube), and R_{target} is a variable component due to power loss in the target. All the separate power losses can be calculated directly, but only if a full finite-element model is used and all regions, including individual conductors in the coil, are meshed.

If the coil is modelled as one 'big' turn, the *skin effect* in the actual wire can be calculated from the following well-known approximation (see exercise 5.14),

(8.2)
$$\frac{R_{\text{a.c.}}}{R_{\text{d.c.}}} = 1 + \frac{1}{48}\left(\frac{a}{\Delta}\right)^4,$$

because the radius of the wire will always be smaller than the skin depth.

The *proximity-effect* calculation can be based on a solution for a conductor with a circular cross section in a transverse field, B. Such a solution may be found in more advanced texts on applied electromagnetism. After some modifications the following formula is recommended

(8.3)
$$R_{\text{prox}} = 0.9\,\frac{\pi^2}{2}\,\frac{n}{S}\,\omega^2 a^4 \sigma \int_s rB^2\,\mathrm{d}s;$$

this involves area integration of the appropriate field component over the coil cross section. Here, n is the number of turns, S is the cross-sectional area of the coil, σ is the conductivity of the wire, and a is the wire radius. The factor 0.9 was determined by numerical experiment and is due to a nonuniform field distribution inside the conductor. This formula has been tested, with remarkable accuracy, against both the full finite-element model and against measurements.

The inductance calculated using the 'cavity' model will be the same for different target materials, because the cavity does not recognize the type of material, only the fact that it is a good conductor. The reistance may be found by integrating the square of the tangential component of the field intensity over the surface of the cavity, as is done, for example, when calculating the power loss in thick plates. Again, this equation will not be derived, but it may be found in specialist books. We can use this equation because we have demonstrated that the field outside the cavity, and on its surface, is the same as for the 'real' case of a good conductor. Thus the equation becomes

(8.4)
$$R_{\text{target}} = \frac{\pi}{i^2 \sigma \Delta} \int rH_{\text{t}}^2\,\mathrm{d}l$$

and surfaces where the field is very weak may be ignored. A similar equation may be used when calculating R_{other} if a cavity model is used.

Typical variations of the coil inductance and resistance with changing distance from the target are shown in Figs 8.11 and 8.12, respectively. L_{coil}

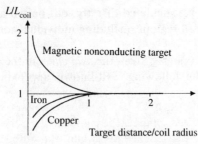

Fig. 8.11 The variation of inductance with the target distance.

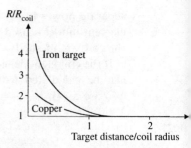

Fig. 8.12 The variation of the resistance with the target distance.

and R_{coil} are the inductance and resistance of the coil when no target is present, and both figures show relative changes of L or R due to bringing the target near. Thus the variation of the reflected impedance of the coil may be predicted with great accuracy.

The example of an inductive sensor has been used here to illustrate the application of electromagnetic theory and computational methods to modelling the performance of a real engineering device. In particular, the practical aspects of such modelling have been discussed, including the various compromises and approximations which we decided to employ. We have demonstrated that electromagnetic analysis not only helps in understanding how devices work but it also gives accurate predictions of performance characteristics and thus it can be used as part of a design process. Many other devices could be analysed in a similar way, and we shall give some more examples in this chapter. The purpose of giving examples is not to provide an encyclopedia of solutions – this would be simply impossible due to the vast number of relevant cases. Our aim is to convince the reader that electromagnetism is a very practical subject and that the combination of theory and appropriate computational tools may be extremely useful.

8.3 The rotating field

An important class of magnetic-field devices are electric machines which operate on the principles of electromechanical energy conversion and thus involve the interchange of energy between electrical and mechanical systems. This concept was introduced in general terms in Section 5.2 and we shall now look at some more practical aspects. When energy is converted from an electrical to a mechanical form, a device is performing a *motor action*. Whereas, a *generator action* involves the conversion of mechanical energy into electrical energy. Electromechanical energy convertors embody three essential features: an electric system, a mechanical system, and a coupling magnetic field. In Section 8.4 we shall investigate the conditions under which an electric circuit must operate to produce an electromagnetic force or torque. In this section we will look into the fundamental property of the magnetic field of most rotating machines, this is called the *rotating-field principle*. We shall demonstrate that a rotating field with a constant amplitude and an approximately sinusoidal space distribution of the m.m.f. around the periphery of the *stator* (that is, the stationary part of the machine) is produced by a three-phase winding located on the stator and excited by balanced three-phase currents when the respective phase windings are wound 120 electrical degrees apart in space. The importance of the above statement lies in the fact that movement (rotation) of the magnetic field may be achieved using stationary system of coils (windings). A typical example of this situation will be found in both of the most common types of a.c. machines, that is, synchronous and induction machines. The *rotor* of these machines rotates either in synchronism with the field (a synchronous machine) or slightly slower to allow induction of currents to happen (an induction machine). In either case we can describe the operation of the machine by saying that the rotor simply 'follows' the rotation of the field.

To understand the rotating-field principle we shall begin by investigating a simple model consisting of three identical concentrated coils located on axes physically 120° apart and supplied from a balanced three-phase supply (that is, with 120° phase difference), as illustrated in Fig. 8.13. In the volume close to the intersection of the three axes, the instantaneous flux densities may be written as

$$B_A = B_m \cos \omega t,$$

$$B_B = B_m \cos(\omega t - 120°),$$

(8.5)
$$B_C = B_m \cos(\omega t - 240°).$$

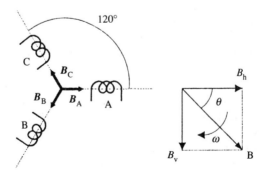

Fig. 8.13 The rotating-field principle.

The resultant flux density produced by this system in the small volume is determined by resolving the components with respect to the physical axes in 'horizontal' and 'vertical' directions. Thus, the sum of the horizontal components gives

$$B_h = B_m[\cos \omega t - \cos(\omega t - 120°)\cos 60° - \cos(\omega t - 240°)\cos 60°]$$

(8.6)
$$= \tfrac{3}{2}B_m \cos \omega t,$$

whereas the sum of the vertical components yields

$$B_v = B_m[0 - \cos(\omega t - 120°)\sin 60° + \cos(\omega t - 240°)\sin 60°]$$

(8.7)
$$= -\tfrac{3}{2}B_m \sin \omega t,$$

The resultant flux is then found as

(8.8)
$$B = (B_v^2 + B_h^2)^{1/2} = \tfrac{3}{2}B_m(\cos^2 \omega t + \sin^2 \omega t)^{1/2} = \tfrac{3}{2}B_m,$$

where $\theta = \omega t$. Hence the resultant flux is synchronous (rotating with angular frequency ω) and of constant magnitude.

In a 'real' electrical machine this model has to be enhanced to account for the actual geometry of the machine where the winding is distributed and is placed in slots in the laminated stator core. Consider the simple two-pole winding arrangement on a stator shown in Fig. 8.14. We can see that the windings of the individual phases are displaced by 120° from each other, in space, around the air-gap periphery and that three subsequent slots are occupied by coils belonging to a particular phase. The winding is single-layered and fully pitched. Many other winding arrangements are possible, and for details the reader is referred to standard texbooks on electrical machines. We are concerned with a fundamental property of the arrangement, and the winding of Fig. 8.14 will be sufficient to demonstrate the principle. The rotor is assumed to be slotted as well (as, for example, in an induction motor), but the effect of the field due to the rotor winding currents will not be included in our discussion.

In order to visualize the rotating-field effect, we shall take a number of 'snapshots' at different instants of time, guided by the time variation of currents in the supplying three-phase system. We therefore assume that at some arbitrary time, $t = 0$, the current in phase A is a maximum, and thus the currents in phases B and C are equal to each other, each being opposite to and half of the current in phase A, so that the algebraic sum of all three currents is always zero. We can use *phasor* notation to demonstrate this position clearly, as shown in Fig. 8.15(a). The magnetic-field distribution may be calculated, and a typical field plot is shown in Fig. 8.15(b). Notice that the field has a definite direction relative to the position of the phase windings.

Magnetic-field distributions at other instants of time may be calculated

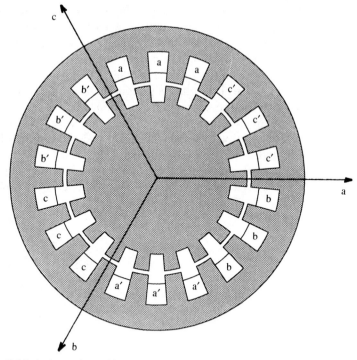

Fig. 8.14 A three-phase distributed winding in a stator.

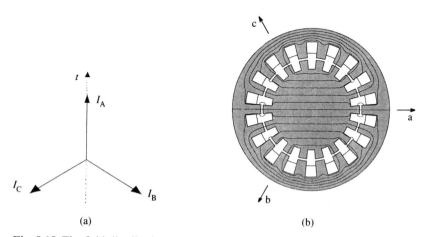

(a)   (b)

Fig. 8.15 The field distribution at $\omega t = 0°$.

in a similar way, and they are shown in Figs 8.16, 8.17, and 8.18 for $\omega t = 30°$, $\omega t = 60°$, and $\omega t = 90°$, respectively. A rotating-time-axis convention has been adopted on the phasor diagrams. Unfortunately, we are not able to produce a moving picture in a book, but comparisons of field plots for different time instants clearly show that as time moves with the sinusoidal waveform of the three-phase currents the direction of the magnetic field follows the same movement in space. For a two-pole case, every angle in the time domain translates directly into a geometrical angle with the same value. For multiple-pole machines the field rotation will be slower by a factor equal to the pole-pair number.

The general conditions for producing a rotating magnetic field are therefore a space distribution of the windings (the geometrical angle

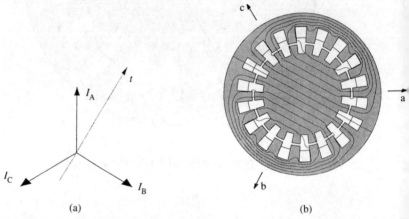

(a)   (b)

Fig. 8.16 The field distribution at $\omega t = 30°$.

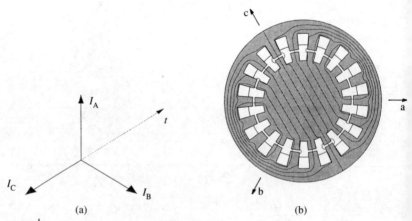

(a)   (b)

Fig. 8.17 The field distribution at $\omega t = 60°$.

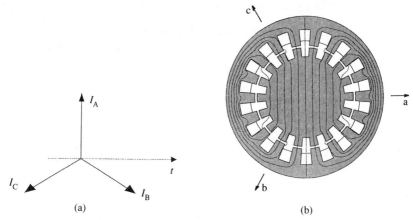

(a) (b)

Fig. 8.18 The field distribution at $\omega t = 90°$.

between coil axes) and an appropriate time difference between component fields (the phase angle between currents). The resultant field will then rotate in synchronism with the supply frequency.

8.4 Torque in rotating machines

Consider the simple model of a doubly excited system of windings shown in Fig. 8.19. Winding 1 is fixed (like the stator of a real machine) and winding 2 is free to rotate (this represents the rotor). Each coil (winding) will have a self-inductance, L_1 and L_2 for winding 1 and 2, respectively; there will also be a mutual inductance between the two windings, say M. All inductances may in general be functions of the angular position of the rotor, θ.

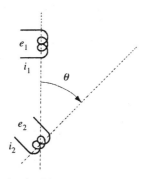

Fig. 8.19 A simple model of a doubly excited machine.

For the system in Fig. 8.19, we can write an equation of energy balance

Mechanical input + electrical input
$$= \text{mechanical storage} + \text{field storage} + \text{losses}.$$

The electrical input is given by

(8.9)
$$dW_{\text{elect}} = (e_1 i_1 + e_2 i_2)\, dt,$$

where

(8.10)
$$e_1 = i_1 R_1 + \frac{d}{dt}(L_1 i_1) + \frac{d}{dt}(M i_2)$$

and

(8.11)
$$e_2 = i_2 R_2 + \frac{d}{dt}(L_2 i_2) + \frac{d}{dt}(M i_1)$$

describe the electric circuits of the two windings. The field storage energy may be expressed in terms of the inductances as

(8.12)
$$W_{\text{fld}} = \tfrac{1}{2} L_1 i_1^2 + \tfrac{1}{2} L_2 i_2^2 + M i_1 i_2,$$

so that

(8.13)
$$dW_{\text{fld}} = \tfrac{1}{2} i_1^2\, dL_1 + \tfrac{1}{2} i_2^2\, dL_2 + i_1 i_2\, dM + (L_1 i_1 + M i_2)\, di_1 + (L_2 i_2 + M i_1)\, di_2.$$

For a very small displacement, the kinetic energy arising from motion will be negligible, whereas the resistance of the windings causes ohmic loss. Collecting the terms and applying the energy-balance equation yields

(8.14)
$$\text{mechanical energy ouput} = \tfrac{1}{2} i_1^2\, dL_1 + \tfrac{1}{2} i_2^2\, dL_2 + i_1 i_2\, dM = T\, d\theta$$

where T is the torque. Hence

(8.15)
$$T = \left(\tfrac{1}{2} i_1^2 \frac{dL_1}{d\theta} + \tfrac{1}{2} i_2^2 \frac{dL_2}{d\theta} \right) + i_1 i_2 \frac{dM}{d\theta}.$$

This equation is known as the *fundamental torque equation* and it is an underlying concept for all electrical machines. It shows that torque will be produced in a rotating device if at least one of the component inductances varies with the rotor position. This may be achieved in a number of ways leading to different types of machines. Such machines will have different characteristics and particular design features, but it is interesting that they all rely on the same principle.

As an example, consider a machine with a uniform air gap, which means that L_1 and L_2 are constants. Let one winding be supplied from a d.c. source and the other from an a.c. source, so that $i_1 = I_1$ and $i_2 = I_2 \cos \omega t$, where I_2 is the peak value of the alternating current. The above conditions apply to a *synchronous machine*. Assume a sinusoidal variation of M with displacement, $M = M_{max} \cos \theta$, so that substitution into the fundamental torque equation, (eqn 8.15), yields

$$T = i_1 i_2 \frac{dM}{d\theta}$$

$$= I_1 I_2 \cos \omega t (-M_{max} \sin \theta)$$

$$= -\tfrac{1}{2} I_1 I_2 M_{max} [\sin(\theta + \omega t) + \sin(\theta - \omega t)].$$

At synchronous speed, $d\theta/dt = \omega$, so that $\theta = (\omega t - \delta)$. Hence

(8.16)
$$T = -\tfrac{1}{2} I_1 I_2 M_{max} [\sin(2\omega t - \delta) + \sin(-\delta)].$$

The average of the time-dependent term is zero. Thus the net average torque

(8.17)
$$T_{av} = \tfrac{1}{2} I_1 I_2 M_{max} \sin \delta.$$

Note that the average torque is proportional to $\sin \delta$, where δ is called the *load angle*.

The *induction-motor* operation will be described by putting sinusoidal a.c. current in both windings, $i_1 = I_1 \cos \omega_1 t$ and $i_2 = I_2 \cos \omega_2 t$, respectively, where I_1 and I_2 are peak values. We shall also assume sinusoidal variation of the mutual inductance with angular displacement, $M = M_{max} \sin \theta$, where $\theta = \omega_3 t$. Note that ω_1, ω_2, and ω_3 are all different. Thus

$$T = (I_1 \cos \omega_2 t)(I_2 \cos \omega_2 t)(M_{max} \cos \omega_3 t).$$

Clearly, various combinations of the time-dependent terms are possible; a condition is sought whereby a term will emerge which is not time-dependent, so that a net average torque is developed. For example, the ω_2 and ω_3 terms can be combined

$$T = \tfrac{1}{2} I_1 I_2 M_{max} \cos \omega_1 t \{\cos[(\omega_2 + \omega_3)t] + \cos[(\omega_2 - \omega_3)t]\},$$

and the condition $\omega_1 = (\omega_2 + \omega_3)$ can be imposed so that

$$T = \tfrac{1}{2} I_1 I_2 M_{max} \{\cos^2 \omega_1 t + \cos \omega_1 t \cos[(\omega_2 - \omega_3)t]\}$$

$$= \tfrac{1}{2} I_1 I_2 M_{max} \{\tfrac{1}{2}(1 + \cos 2\omega_1 t) + \cos \omega_1 t \cos[(\omega_2 - \omega_3)t]\}.$$

Thus the net average torque

(8.18)
$$T_{av} = \tfrac{1}{4} I_1 I_2 M_{max}.$$

The condition chosen for the three angular frequencies may be stated as $\omega_3 = (\omega_1 - \omega_2)$, that is, the rotational speed is directly related to the frequencies of the currents. This condition demonstrates that torque is developed when $\omega_2 = (\omega_1 - \omega_3)$, where ω_2 is termed the *slip speed* and is given by the difference between the speed of the field produced by the stator winding and the speed of rotation of the rotor. The concept of slip is of fundamental importance in studies of induction motors.

Finally, consider a singly excited *reluctance motor*, which has only one winding (on the stator), but the rotor is magnetically nonsymmetrical so that the self-inductance of the stator winding is a function of position. Let $L_1 = L_0 + L \cos 2\theta$ and $i_1 = I \cos \omega t$, so that

$$T = -LI^2 \cos^2 \omega t \sin 2\theta$$

$$= -\tfrac{1}{2}LI^2(\sin 2\theta + \sin 2\theta \cos 2\omega t)$$

$$= -\tfrac{1}{2}LI^2(\sin 2\theta + \tfrac{1}{2}\sin[2\theta + 2\omega t] + \tfrac{1}{2}\sin[2\theta - 2\omega t]).$$

Under normal conditions of operation, $\theta = (\omega t - \delta)$, and thus $d\theta/dt = \omega$, so that the running speed is synchronous. Then

$$T = -\tfrac{1}{2}LI^2[\sin(2\omega t - 2\delta) + \tfrac{1}{2}\sin(4\omega t - 2\delta) + \tfrac{1}{2}\sin(-2\delta)]$$

and

(8.19)
$$T_{av} = \tfrac{1}{4}LI^2 \sin 2\delta,$$

where δ is the load angle. Hence, if the machine has a nonuniform air gap, as in a machine with 'salient poles', the rotor rotates with synchronous speed due to the reluctance torque. Note that a machine with salient poles and double excitation develops both reluctance and excitation torques when running at synchronous speed.

8.5 Energy storage and forces in magnetic-field systems

The energy associated with a magnetic field is distributed throughout the space occupied by the field. Assuming no losses and linear magnetic characteristics (constant permeability), the magnetic energy density is given by

(8.20)
$$w_m = \tfrac{1}{2}BH = \tfrac{1}{2}\frac{B^2}{\mu_r \mu_0},$$

its units are joules per cubic metre (J m^{-3}). The greater the value of the relative permeability, the less is the energy stored for a given value of flux density. Clearly, the energy stored in an air gap may be several times that stored in a much greater volume of iron. In practical electromagnetic devices built with air gaps, the magnetic nonlinearity and the core losses may often be neglected, and a linear analysis may be justified.

For a linear region such as an air gap, the energy stored can be expressed as

(8.21)
$$W_{m\,air} = \frac{1}{2}\left(\frac{B^2}{\mu_0}\right) \times \text{volume} = \tfrac{1}{2}\,\text{m.m.f.} \times \phi = \tfrac{1}{2}R_\mu \phi^2 = \tfrac{1}{2}i\lambda,$$

where R_μ is the reluctance of the air gap and λ is the flux linkage in weber-turns given by the product of a number of turns, n, and the flux, ϕ, linking n. This equation expresses the amount of energy stored upon increasing the flux density from zero to B. Most of the energy is required to establish the flux in the air gap.

A change in flux density from a value of zero initial flux density to a value B requires an energy input to the field occupying a given volume of

(8.22)
$$W_m = \text{volume} \times \int_0^B H \, dB$$

which for a singly excited system can also be expressed by

(8.23)
$$W_m = \int_0^\lambda i(\lambda) \, d\lambda = \int_0^\phi \text{m.m.f.}(\phi) \, d\phi.$$

Note that the current is a function of the flux linkages, and the m.m.f. is a function of the flux; their relations depend on the geometry of the magnetic circuit and windings, as well as the magnetic properties of the core material. Equations (8.22) and (8.23) may be interpreted graphically as the area labelled as *energy* in Fig. 8.20. The other area, labelled as *coenergy*, can be expressed as

(8.24)
$$W'_m = \text{volume} \times \int_0^H B \, dH = \int_0^i \lambda(i) \, di = \int_0^{\text{m.m.f.}} \phi(\text{m.m.f.}) \, d(\text{m.m.f.})$$

For a linear system in which B and H, or λ and i, or ϕ and m.m.f., are proportional, the energy and coenergy are equal. However, for a nonlinear system the energy and coenergy will differ, but their sum for a singly excited system is given by

(8.25)
$$W'_m + W_m = \text{volume} \times BH = \lambda i = \text{m.m.f.} \ \phi.$$

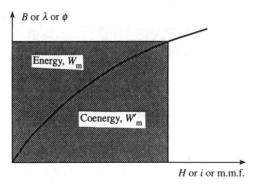

Fig. 8.20 The energy and coenergy.

Consider a simple electromechanical system where movement is allowed in one direction only, say x, and electrical and mechanical losses are neglected. An energy balance equation may be written as

Electrical energy input
$\qquad$ = mechanical energy output + increase in stored energy.

Figure 8.21 illustrates a change at constant λ when the x-coordinate is changed. Since there is no electrical input at constant flux then

$$F = -\left(\frac{\partial W_{\mathrm{m}}}{\partial x}\right)_{\lambda\,=\,\mathrm{constant}}.$$

(8.26)

Now consider Fig. 8.22 which shows a change of W_{m} at constant current and the associated electrical input due to the changing flux. Now

$$i\left(\frac{\partial \lambda}{\partial x}\right)_{i\,=\,\mathrm{constant}} - \left(\frac{\partial W_{\mathrm{m}}}{\partial x}\right)_{i\,=\,\mathrm{constant}} = F,$$

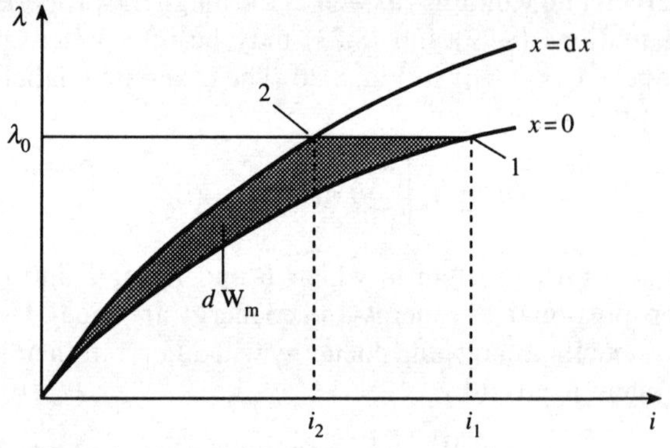

(8.27)

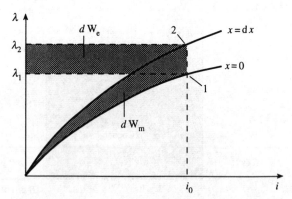

Fig. 8.21 The change in the stored energy at a constant flux.

Fig. 8.22 The change in the energy at a constant current.

but the left-hand side can be written in terms of the coenergy because

(8.28)
$$i\left(\frac{\partial \lambda}{\partial x}\right)_{i = \text{constant}} = \left(\frac{\partial (i\lambda)}{\partial x}\right)_{i = \text{constant}}$$

and

(8.29)
$$\left(\frac{\partial (i\lambda)}{\partial x}\right)_{i = \text{constant}} - \left(\frac{\partial W_m}{\partial x}\right)_{i = \text{constant}} = \left(\frac{\partial W'_m}{\partial x}\right)_{i = \text{constant}},$$

so that F can be written in terms of the coenergy as

(8.30)
$$F = +\left(\frac{\partial W'_m}{\partial x}\right)_{i = \text{constant}}.$$

The change in sign between eqns (8.26) and (8.30) should be noted. If the relationship between λ and i is a straight line the magnitude of the two expressions for the force is the same but the sign is reversed because of the electrical input.

This straight-line relationship means that the inductance, L, is a constant. We can write

(8.31)
$$F = +\frac{\partial}{\partial x}\left(\tfrac{1}{2}Li^2\right)_{i = \text{constant}} = \tfrac{1}{2}i^2\frac{\partial L}{\partial x}.$$

The force therefore acts in the direction which increases the inductance. This equation was encountered in Example 3.9.

The calculations of forces and torques based on the *energy method* described above are usually quite convenient. They involve volume integration of the energy or coenergy, and such routines are usually provided by postprocessors of commercial software. However, at least two field solutions are required in order to determine the space derivatives. For improved accuracy, several solutions may be necessary, and thus the computation times may be long. Alternatively, the forces may be calculated using eqn (5.11), but only when the flux density is known and the geometry of the system is rather simple. Yet another approach, known as the *Maxwell stress method*, uses the tensile and compressive stresses along iron surfaces and relies on surface integration of appropriate expressions. Stresses in the magnetic field were discussed in Section 3.11.

8.6 Magnetic materials

In the discussion of magnetic circuits in Section 3.6 the relative permeability, μ_r, of iron was treated as a constant, and a typical value was $\mu_r = 3000$. This is a useful simplification, but one that has to be used with caution. The behaviour of magnetic materials is varied and complicated, and different materials are used for different technological purposes. A helpful survey of such properties is given in the book *Lectures on the electrical properties of materials* by L. Solymar and D. Walsh (Oxford University Press, 5th edn, 1993). In this section we shall give a very brief and basic summary.

One of the important features of atomic structure is *electron spin* which associates a magnetic moment with the electrons in a way similar to the equivalence between a small current loop and a magnetic dipole which is illustrated in Fig. 3.2. In ferromagnetic materials (iron, nickel, and cobalt), the spins of electrons and the magnetic moments add up to give a bulk magnetization to a region called a domain. If all the domains were to be aligned, the bulk material would have a strong external field which would be associated with a large amount of field energy and a strong surface polarity as illustrated in Fig. 8.23. The surface polarity gives an internal magnetic field strength opposing the flux density, **B**. The crystalline structure of iron produces preferred directions for the domain magnetization along the sides, and to a lesser extent along the diagonals of the cubic crystals. The preferred orientations can be in either direction, and a more likely domain arrangement, in which the field energy is reduced, is given by Fig. 8.24.

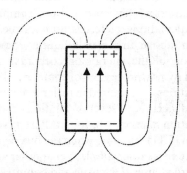

Fig. 8.23 The field of a strong magnet.

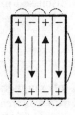

Fig. 8.24 A weak magnetic field.

The subdivision of the domains is limited by the fact that the direction of the magnetization has to change at the boundaries between domains. This involves additional energy. Moreover, local impurities will be associated with the surface polarity which provides other constraints and causes the domain walls to stick to the impurities. The final minimum-energy position therefore depends on a large variety of factors. In iron which has not been magnetized by external fields, the domain effects cancel and the bulk material appears to be nonmagnetic, although it consists of an assembly of strongly magnetized small regions.

Consider what happens when such an unmagnetized piece of iron is subjected to a magnetic field. In order to simplify the arrangement we shall examine an iron ring of constant cross section on which is wound a uniformly distributed multiturn coil carrying a current. In this arrangement, the surface polarity is negligible and the magnetic field strength, H, is entirely due to the current. The average flux density, B, in the iron can be measured by means of a search coil wound on the ring, which is connected to a flux meter. Since H is proportional to the magnetizing current i, we can obtain the B–H curve for the iron. Such a curve is typified by Fig. 8.25. The curve exhibits three regions which have been labelled as a, b, and c. In terms of domain structure, the regions can be explained as follows. At first the domains cancel each other, as shown in Fig. 8.26. As the applied field H is increased B increases due to the motion of the domain walls, which favours

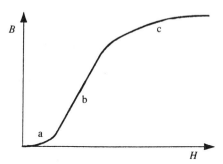

Fig. 8.25 A typical B–H curve for iron.

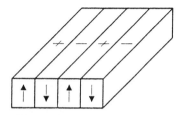

Fig. 8.26 Domains cancelling each other.

those domains in the direction of **H**, as shown in Fig. 8.27. This behaviour describes the foot, a, of the curve in Fig. 8.25. As **H** is further increased, the domains which are in the opposing direction reverse, causing a rapid increase of **B** and giving a relationship of **B** to **H** which is approximately linear. The domains are then more or less in the same direction, as shown in Fig. 8.28. The increase of **B** with increasing H then becomes very much smaller, as shown in region c in Fig. 8.25. The domains are now parallel with **H**, as shown in Fig. 8.29. The iron is now said to approach saturation. Since $B = \mu_0 \mu_r H$, the relative permeability approaches unity and the B–H curve becomes almost horizontal. Clearly μ_r varies widely, but for many applications the foot, a, in Fig. 8.25 is small and the saturated region, c, is

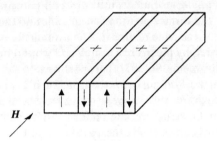

Fig. 8.27 More domains in the direction of applied field.

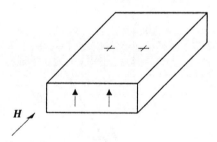

Fig. 8.28 All domains in the same direction.

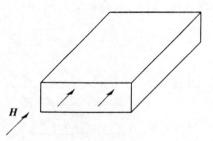

Fig. 8.29 Domains parallel to the applied field.

not used because it requires large values of the magnetizing current and therefore large ohmic losses. Hence the region b in Fig. 8.25 is the most important, and the use of a constant value for μ_r provides a reasonable approximation. In more accurate design calculations, different values of μ_r are used for the different parts of devices.

There is, however, a further complication which is of great engineering importance. The movement of domain walls and the reversal of domains give rise to local e.m.f.s and local *eddy currents* in the conducting material. Hence there are energy losses associated with changes in the magnetization. Moreover, there are friction losses caused by impurities which distort the crystal structure and the domains. This means that the *B–H* curve is not single valued. When the current is reduced, the flux density does not follow the path of Fig. 8.25. Instead we obtain the hysteresis curve in Fig. 8.30. The energy loss is $\oint B \, dH$ around the loop and it is therefore equal to its area. Small excursions of *H* give rise to minor loops such as the one illustrated.

Hysteresis loops are a disadvantage in some applications, but they can be put to good use in others. In transformers, for example, the loss is undesirable because it wastes energy as heat, which has to be removed by cooling. The best transformer iron has a thin hysteresis loop. Such a material is said to be magnetically soft because it allows easy movement to the domain walls. Moreover, the conductivity is reduced as far as possible in order to inhibit the eddy currents associated with the domains. It has been found that small amounts of silicon reduce hysteresis effects. In soft iron hysteresis loops still exists, but they lie close to the single-valued *B–H* curve of Fig. 8.25.

In many applications a constant magnetic field is required to produce a force or a torque. Often this can be done by using permanent magnets made from 'hard' magnetic materials having a wide hysteresis loop. Such materials contain various substances which impede domain movement. During the manufacture of these materials special cooling processes are

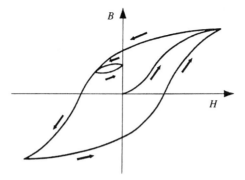

Fig. 8.30 Hysteresis loops.

used to freeze the domains into position as far as possible. The region of the hysteresis curve of special interest for permanent magnets is that in which B is positive and H is negative. This is the condition which enables magnets to sustain a negative field strength internally while providing an external field. There is then no need to provide a magnetizing winding with its additional initial and running costs. A flat B–H curve is very desirable, so that large amounts of surface polarity and negative magnetic field strength can be used. Figure 8.31 shows some typical B–H curves for modern permanent-magnet materials. The value of negative H at zero B is called the *coercive force*. Typical values are $H = 10^5$–10^6 A m^{-1} for permanent magnets which can be compared with $H = 10^2$ A m^{-1} for a soft silicon iron. The flux densities of the two types of material are similar, as would be expected from their domain structure. Another important class of magnetic materials for applications at high frequencies are the ferrites, which are various iron-oxide combinations. Unlike metallic iron these materials are insulators and not conductors. This eliminates their eddy-current losses almost entirely, and it also reduces their hysteresis losses. On the other hand, their flux densities and permeabilities are considerably less than those of iron.

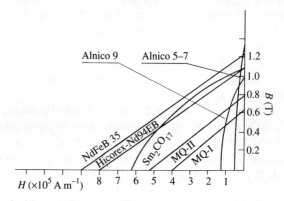

Fig. 8.31 Typical hysteresis curves of permanent-magnet materials.

Appendix 1

TAS tutorial

The system under consideration for calculation of the resistance between a pair of electrodes is shown in Fig. A1.1.

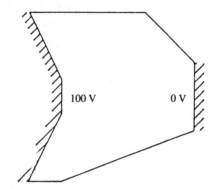

Fig. A1.1 Calculation of the resistance between a pair of electrodes.

When you run the TAS program you are put into the *Welcome* screen menu (see Fig. A1.2). Different options are available by moving the mouse so that the appropriate box is highlighted. If this is the first time you run the program you are encouraged to go through the *Introduction* and the *Example*. Otherwise choose the *Input Screen* option and press the *left-hand button* (LHB) or the *right-hand button* (RHB) on the mouse.

You are now in the *Input Screen* menu, see Fig. A1.3. Notice that communication with other menus is provided by boxes below the TAS logo (*Main Screen* and *Welcome*). You can move between the different menu screens at any time without losing data. Also notice the *Help* and *Hint* options (to the left of the TAS logo) – try using them. Now, move the mouse so that the marker (arrow) is inside the *Automatic tube/slice generator* box and press the LHR or the RHB (the box at the bottom right-hand corner of the screen always tells you what action to expect if one or both mouse buttons are pressed). A *Help* screen will appear with some explanatory notes. Read the message and press the mouse button to continue. You are now in the *Auto t/s generator* subprogram.

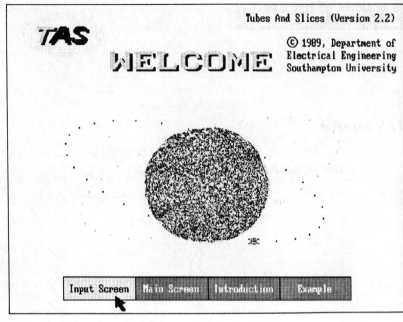

Fig. A1.2 The *Welcome* screen.

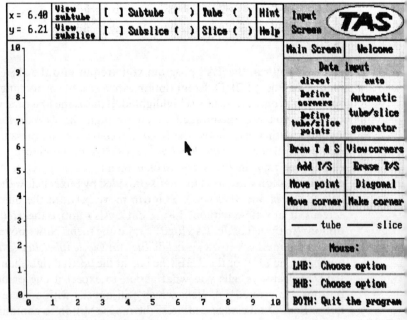

Fig. A1.3 The *Input* screen.

Move the mouse to the position $x = 0.00$ and $y = 0.00$ (watch the top left-hand corner of the screen where the coordinates are displayed) and press the LHB. This will create a point at the specified position and at the same time it will define a *tube* line. As you now move the mouse a white elastic line follows the marker movements (see Fig. A1.4). Position the marker at the point $x = 2.00$ and $y = 0.00$ and press the LHB. You have created a tube boundary between the first two points. The next boundary line is also defined as a tube. Notice that pressing the RHB creates a red line depicting a *slice* boundary.

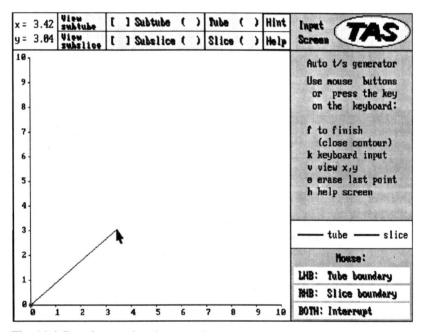

Fig. A1.4 Data input using the *Auto t/s generator* option.

Input the remaining *corner* points in the following sequence:

(1) move to $x = 10.00$, $y = 3.00$, and press the RHB;
(2) move to $x = 10.00$, $y = 7.00$, and press the LHB;
(3) move to $x = 7.00$, $y = 10.00$, and press the LHB;
(4) move to $x = 0.00$, $y = 10.00$, and press the RHB;
(5) move to $x = 2.00$, $y = 6.00$, and press the RHB; and
(6) move to $x = 2.00$, $y = 4.00$, and press the RHB.

The screen should now appear as shown in Fig. A1.5. **Do not** input another point at $x = 0.00$, $y = 0.00$, since all the necessary information has already

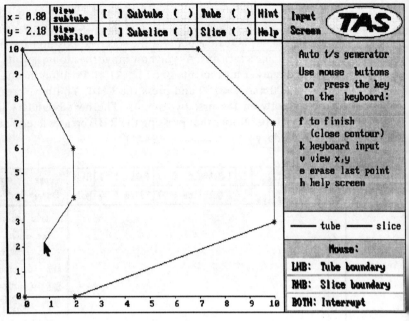

Fig. A1.5 The position of all *corner* points.

been supplied. If you make a mistake press the letter e on the keyboard to erase the most recent point. If you prefer to enter coordinates from the keyboard press the letter k, then type in *x*- and *y*-values, as requested, and finally press the letter s or t to define a *slice* or a *tube* line. Other options include pressing the letter v to view the coordinates of defined points (that is, display their values), and the letter h to display the *help* screen.

All *corners* and boundary lines have now been defined. Press the letter f to finish the data input. The *Input Screen* menu reappears. Choose the *View corners* option to check the corner-points coordinates (see Fig. A1.6). At the same time inspect the distribution of *tube/slice* lines. These lines have been generated automatically on the basis of the shape of boundary lines, but these are not necessarily in the correct position. Press the mouse button to continue.

The data is now complete, even if the assumed distribution is not optimal. Point the arrow at the *Main Screen* box and press the mouse button. The *Main Screen* menu will appear. The final distribution of *subtube/subslice* lines is shown. In order to find the value of the resistance move the cursor inside the *Resistance* box and press the mouse button. The upper and lower bounds of the resistance, as well as the average value, are displayed instantly. The percentage error (confidence limits) of the solution is also shown. Compare your results with Fig. A1.7.

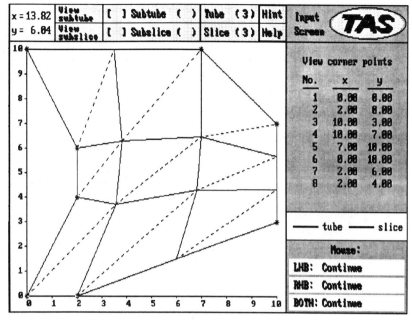

Fig. A1.6 The *View corners* option.

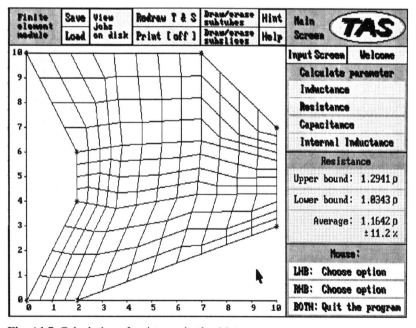

Fig. A1.7 Calculation of resistance in the *Main screen* menu.

Go back to the *Input Screen* menu and choose the *Move point* option. A different cursor appears on the screen. Position the cursor close to the point $x = 3.83$, $y = 6.29$ and press the LHB. Press the LHB again and while **keeping it pressed** move the point to a new position $x = 3.80$, $y = 6.07$. Release the LHB and press the RHB. Reposition the other points:

(1) point (3.58, 3.71) to (3.50, 3.82);
(2) point (6.82, 4.29) to (6.18, 3.89); and
(3) point (7.00, 6.46) to (6.28, 5.86).

Notice that not only the interior points but also those boundary points which are not *corners* may be repositioned using the *Move point* option. Inspect the new tube/slice distribution shown in Fig. A1.8.

Press the RHB to leave the *Move point* option. Go to the *Main Screen* menu and choose the *Print [ ]* option. Press **both** mouse buttons to switch to an automatic printout of results. Make sure your printer is connected and switched on. Calculate the resistance again by pointing the cursor at the *Resistance* box and pressing the mouse button.

Your screen should be similar to the screen shown in Fig. A1.9.

At the same time as the calculation of resistance is performed, a full printout of data and results is provided. If your printer is properly connected and switched on, your printout should be similar to that shown

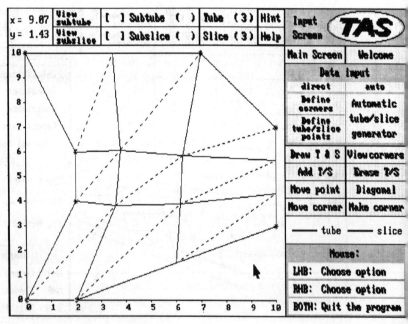

Fig. A1.8 The *Move point* option.

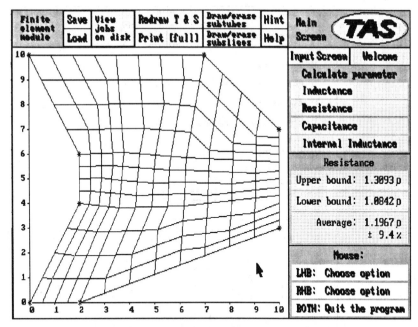

Fig. A1.9 Calculation using the *Print* [*full*] option.

in Fig. A1.10. Notice that a short version of a printout (the results only) is also possible.

Before proceeding any further, compare the two results for resistance that you have obtained with the two tube/slice distributions, and notice the improvement in the results (a reduction of the percentage error in the average value).

You are encouraged to explore the other options available in the *Input Screen* menu, in order to achieve even better accuracy of computation. In particular, you may try a combination of the *Add T/S*, *Diagonal* and *Move point* options. Remember that your aim is to produce an approximately orthogonal system of tube/slice lines.

We shall concentrate now on the use of the *finite element* extension to the TAS program.

Choose the *Finite element module* in the top left corner of the *Main Screen* menu. A new screen menu will appear for preprocessing of the finite-element method (FEM). As you can see in Fig. A1.11, the finite-element mesh is based directly on the tube/slice distribution from the *Input Screen*, and it gives 18 elements and 16 nodes. In order to achieve sufficient accuracy of the FEM solution, some mesh refinement will be advisable. Point the marker at the box *Tube (3)* and press the LHB. Press the LHB again. You have doubled the number of tubes.

```
TAS results
===========

Resistance:

Upper bound:   R+   =  1.3093

Lower bound:   R_   =  1.0842

    Average:   Rave =  1.1967


Full printout of data
---------------------

number of fixed points =  8
        number of tubes =  3
       number of slices =  3

------------------------------------------------------------------
             position   which      point    number of number of
  m   n       x      y  diagonal    type     subtubes  subslices
 ---  ---    ----  ----  --------   --------  --------- ---------

  0   0     0.00  0.00     /        corner       4         4
  1   0     2.00  0.00     /        corner       4         4
  2   0     6.00  1.50     /        boundary     4         4
  3   0    10.00  3.00              corner

  0   1     2.00  4.00     /        corner       4         4
  1   1     3.60  3.82     /        internal     4         4
  2   1     6.18  3.89     /        internal     4         4
  3   1    10.00  4.32              boundary

  0   2     2.00  6.00     /        corner       4         4
  1   2     3.80  6.07     /        internal     4         4
  2   2     6.28  5.86     /        internal     4         4
  3   2    10.00  5.68              boundary

  0   3     0.00 10.00              corner
  1   3     3.50 10.00              boundary
  2   3     7.00 10.00              corner
  3   3    10.00  7.00              corner

------------------------------------------------------------------
Note: / means diagonal between points [m][n] and [m+1][n+1]
      \ means diagonal between points [m][n+1] and [m+1][n]
      number of subtubes/subslices refers to subdivisions
      between [m][n] and [m][n+1] / [m][n] and [m+1][n]
```

Fig. A1.10 A printout of the results.

Double the number of tubes again. Now double the number of slices twice. Watch the number of elements and number of nodes as you keep refining the mesh. At this stage you have created a mesh of 12 tubes and 12 slices, hence there are 288 elements and 169 nodes (see Fig. A1.12). Try using the *Add T/S* option as an alternative mesh-refinement procedure. There is a limit of 25 tubes and 25 slices which you may generate for this purpose.

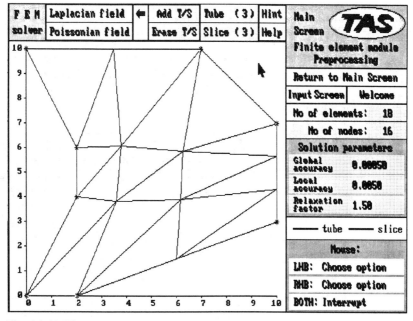

Fig. A1.11 The *Finite element module.*

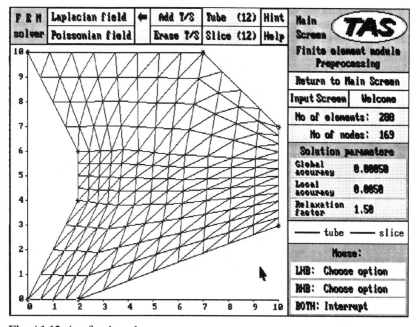

Fig. A1.12 A refined mesh.

Choose the *FEM solver* option. After a few seconds taken by the solution you will be put into postprocessing mode. Notice that the problem is solved twice to accommodate the dual formulation. Choose the *Resistance* option in the *Calculate parameter* window (see Fig. A1.13) to display the value of resistance. Compare the results with the values calculated previously in the *Main Screen* menu. In particular, notice that the average value is hardly changed.

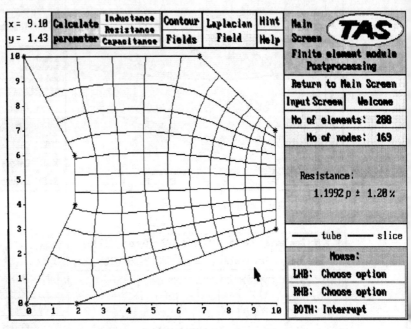

Fig. A1.13 Postprocessing in the *Finite element module*.

The contour lines are permanently displayed. These are essentially tubes and slices provided by the finite-element solution, and for a steady current flow they represent current flow and equipotentials. Try the *Fields* option. Position the cross-shaped marker at point $x = 4.18$, $y = 5.75$, and press the LHB. The value of the potential and its derivatives at that point will be shown (see Fig. A1.14). Find the potential at other internal and boundary points. Press the RHB to abandon this option. Press **both** buttons (or choose the *Return to Main Screen* option) to go back to the *Main Screen* menu.

In order to use the alternative *direct* (that is, manual) input mode, go to the *Input Screen* menu and choose the *Define corners* option. Define the position of all *corner* points, using a mouse or from the keyboard, in a

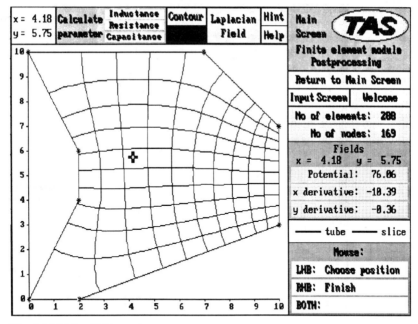

Fig. A1.14 The *Fields* option.

similar way as used in the *auto* data input, only this time you need not specify the type of line. Next, choose the option *Tube ( )* and enter the number of tubes, say 5. Choose the *Slice ( )* option and enter number of slices, say 4. Finally, choose the *Define tube/slice points* option.

You now enter, by pressing the LHB, coordinates of all 30 internal and boundary points, starting with the point $x = 0.00$, $y = 0.00$, and going *along* tubes lines. To define a point at a *corner* or on a *boundary line* move the marker close to the required position, but not necessarily to the precise location, and press the LHB. Internal points are created at the exact position of the marker. A sample screen from the intermediate stage of *Construction points* input is shown in Fig. A1.15.

After you have defined the last point at the position $x = 10.00$, $y = 7.00$, your screen may look similar to that shown in Fig. A1.16.

Go to the *Main Screen* and calculate the resistance. Compare results with previously obtained values (see Fig. A1.17).

Go back to the *Input Screen* menu and choose the *Move corner* option. Pick up the *corner* point $x = 10.00$, $y = 3.00$, and move it (keep the LHB pressed!) to a new position $x = 10.00$, $y = 1.00$ (see Fig. A1.18). Watch how the tube/slice lines follow the changes of the boundary shape. Press the RHB to abandon the chosen corner, and press RHB again to return to *Input Screen*.

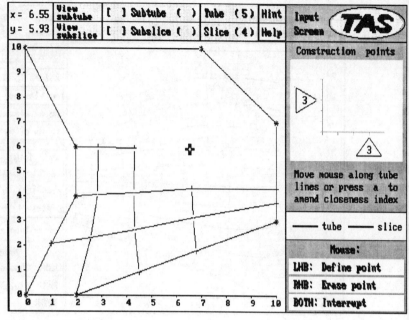

Fig. A1.15 The *Construction points* input.

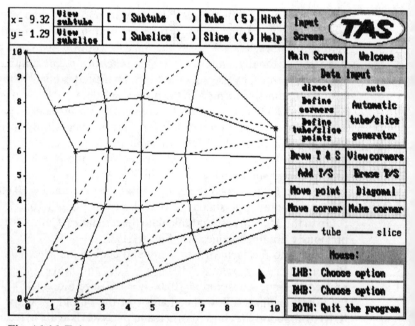

Fig. A1.16 Tubes and slices generated manually.

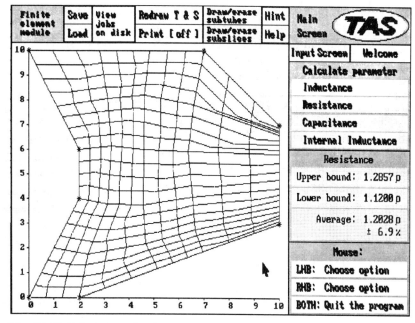

Fig. A1.17 Calculation of resistance.

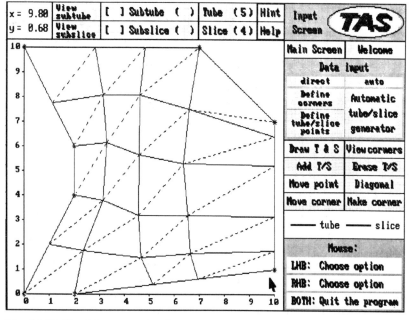

Fig. A1.18 The *Move corner* option.

You have just created a new system, but owing to a minor change made to the original shape you have been able to preserve the basic tube/slice distribution. This may be useful when optimizing shapes.

You can go straight to the *Main Screen* menu to calculate the resistance of the modified geometry (see Fig. A1.19).

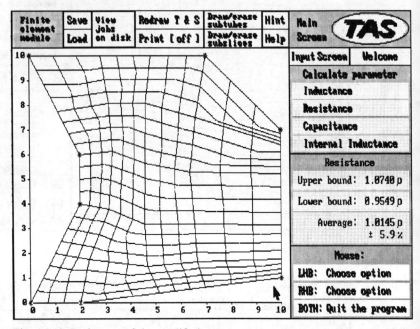

Fig. A1.19 Resistance of the modified system.

Throughout the TAS program, various *error checking* procedures are active. In order to demonstrate their performance let us make a deliberate mistake in data input. Choose the *Move point* option in the *Input Screen* menu and move one of the internal points in such a way that some tube/slice lines are crossing over (see the example in Fig. A1.20). You have created an impossible tube/slice distribution.

When you try to link to the *Main Screen* menu an error message will be displayed (see Fig. A1.21). Notice that no calculations are permitted until you make the necessary corrections to the distribution using the various options available in the *Input Screen*.

In order to *save, load,* or *view* data files on a floppy disk in drive A:\ use appropriate options in the *Main Screen* menu. A sample menu, displayed when the *Load* option is used, is shown in Fig. A1.22.

Finally, remember about the *Help* and *Hint* screens which are available in all screen menus (see Fig. A1.23). Information is provided for all options. The original data is always recovered once the *HELP* facility is abandoned.

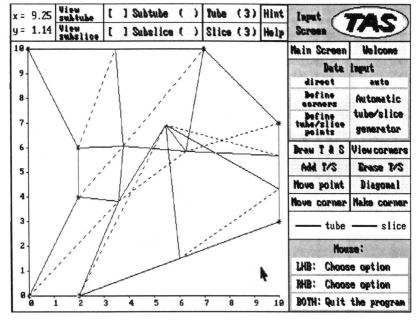

Fig. A1.20 Crossing of tube/slice lines.

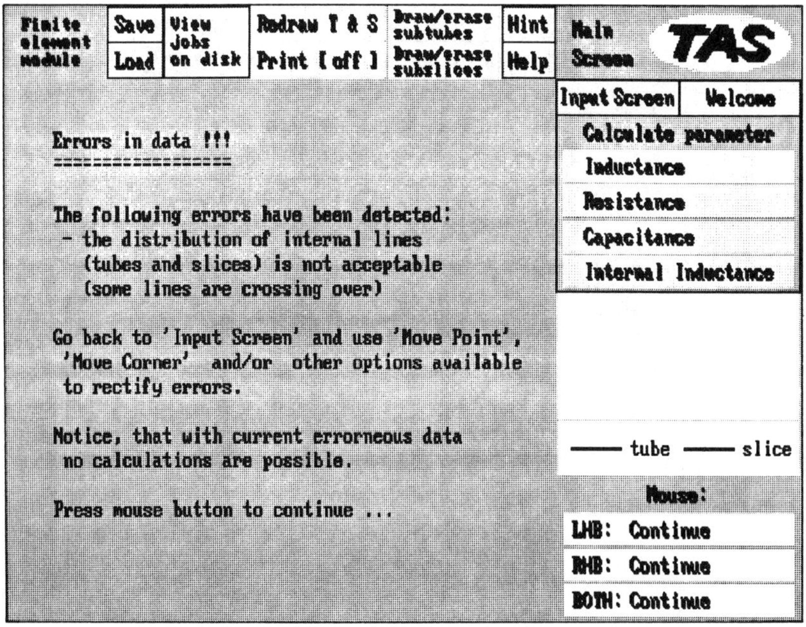

Fig. A1.21 A display of errors.

Finite element module
Save
View Jobs on disk
Redraw T & S
Print [off]
Draw/erase subtubes
Draw/erase subslices
Hint
Help
Main Screen
TAS
Input Screen
Welcome
Load job
Page: 1
No. Date Description
1 11/3/1993 Example 1
2 11/3/1993 Example 2
3 11/3/1993 Example 3
4 11/3/1993 Example 4
5 empty
6 empty
7 empty
8 empty
9 empty
Enter position (1 - 9) from which you want the data to be loaded.
Press LHB to see nest page.
Press RHB to see previous page.
Press BOTH to interrupt.
Calculate parameter
Inductance
Resistance
Capacitance
Internal Inductance
tube slice
Mouse:
LHB: Next page
RHB: Previous page
BOTH: Interrupt

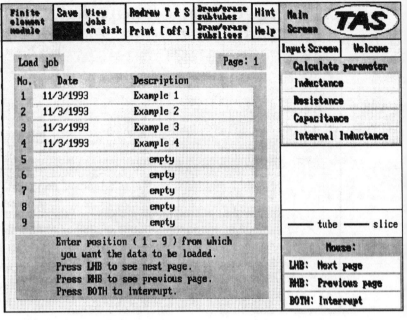

Fig. A1.22 The *Load* option.

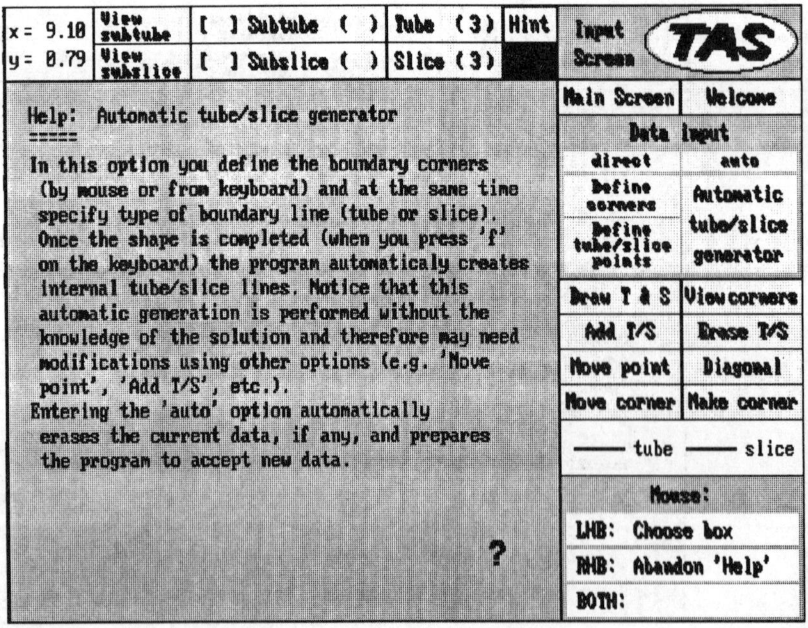

x = 9.10
y = 0.79
View subtube
View subslice
[ ] Subtube ( )
[ ] Subslice ( )
Tube (3)
Slice (3)
Hint
Input Screen
TAS
Main Screen
Welcome
Help: Automatic tube/slice generator
In this option you define the boundary corners (by mouse or from keyboard) and at the same time specify type of boundary line (tube or slice). Once the shape is completed (when you press 'f' on the keyboard) the program automaticaly creates internal tube/slice lines. Notice that this automatic generation is performed without the knowledge of the solution and therefore may need modifications using other options (e.g. 'Move point', 'Add T/S', etc.).
Entering the 'auto' option automatically erases the current data, if any, and prepares the program to accept new data.
?
Data input
direct
auto
Define corners
Define tube/slice points
Automatic tube/slice generator
Draw T & S
View corners
Add T/S
Erase T/S
Move point
Diagonal
Move corner
Make corner
tube slice
Mouse:
LHB: Choose box
RHB: Abandon 'Help'
BOTH:

Fig. A1.23 The *Help* screen.

Appendix 2: Useful mathematical formulae

A2.1 Three common coordinate systems

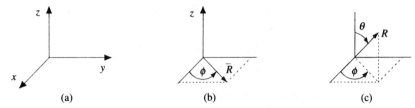

Fig. A2.1 (a) Rectangular coordinates, (b) cylindrical coordinates, and (c) spherical coordinates.

A2.2 Vector addition and multiplication

$$F + G = G + F,$$

$$F \cdot G = G \cdot F,$$

$$F \cdot F = F^2,$$

$$F \times G = -G \times F,$$

$$(F + G) \cdot H = F \cdot H + G \cdot H,$$

$$(F + G) \times H = F \times H + G \times H,$$

$$F \cdot G \times H = G \cdot H \times F = H \cdot F \times G,$$

$$F \times (G \times H) = (F \cdot H)G - (F \cdot G)H.$$

In rectangular coordinates

$$F + G = (F_x + G_x)\hat{x} + (F_y + G_y)\hat{y} + (F_z + G_z)z,$$

$$F \cdot G = F_x G_x + F_y G_y + F_z G_z,$$

$$F \times G = \begin{vmatrix} \hat{x} & \hat{y} & \hat{z} \\ F_x & F_y & F_z \\ G_x & G_y & G_z \end{vmatrix}.$$

A2.3 Gradient, divergence, curl, and Laplacian

$$\text{grad } V = \nabla V = \hat{u}_l \frac{dV}{dl} \quad \text{(maximum slope)}$$

$$\text{div } \boldsymbol{F} = \nabla \cdot \boldsymbol{F} = \lim_{v \to 0} \frac{1}{v} \oint \boldsymbol{F} \cdot d\boldsymbol{s} \quad \text{(outflow per unit volume)}$$

$$\text{curl } \boldsymbol{F} = \nabla \times \boldsymbol{F} = \lim_{s \to 0} \frac{\hat{u}_n}{s} \oint \boldsymbol{F} \cdot d\boldsymbol{l} \quad \text{(circulation per unit area)}$$

$$\nabla^2 \boldsymbol{F} = \text{grad}(\text{div } \boldsymbol{F}) - \text{curl}(\text{curl } \boldsymbol{F})$$

$$\nabla^2 V = \text{div}(\text{grad } V) = \nabla \cdot \nabla V$$

In rectangular coordinates

$$\nabla = \frac{\partial}{\partial x} \hat{x} + \frac{\partial}{\partial y} \hat{y} + \frac{\partial}{\partial z} \hat{z}$$

Hence

$$\nabla V = \frac{\partial V}{\partial x} \hat{x} + \frac{\partial V}{\partial y} \hat{y} + \frac{\partial V}{\partial z} \hat{z},$$

$$\nabla \cdot \boldsymbol{F} = \frac{\partial F_x}{\partial x} + \frac{\partial F_y}{\partial y} + \frac{\partial F_z}{\partial z},$$

$$\nabla \times \boldsymbol{F} = \begin{vmatrix} \hat{x} & \hat{y} & \hat{z} \\ \dfrac{\partial}{\partial x} & \dfrac{\partial}{\partial y} & \dfrac{\partial}{\partial z} \\ F_x & F_y & F_z \end{vmatrix},$$

or

$$\nabla \times \boldsymbol{F} = \left(\frac{\partial F_z}{\partial y} - \frac{\partial F_y}{\partial z} \right) \hat{x} + \left(\frac{\partial F_x}{\partial z} - \frac{\partial F_z}{\partial x} \right) \hat{y} + \left(\frac{\partial F_y}{\partial x} - \frac{\partial F_x}{\partial y} \right) \hat{z},$$

$$\nabla^2 V = \frac{\partial^2 V}{\partial x^2} + \frac{\partial^2 V}{\partial y^2} + \frac{\partial^2 V}{\partial z^2},$$

$$\nabla^2 \boldsymbol{F} = \nabla^2 F_x \hat{x} + \nabla^2 F_y \hat{y} + \nabla^2 F_z \hat{z}.$$

In cylindrical coordinates

$$\nabla V = \frac{\partial V}{\partial R}\hat{R} + \frac{1}{R}\frac{\partial V}{\partial \phi}\hat{\phi} + \frac{\partial V}{\partial z}\hat{z},$$

$$\nabla \cdot F = \frac{1}{R}\frac{\partial(RF_R)}{\partial R} + \frac{1}{R}\frac{\partial F_\phi}{\partial \phi} + \frac{\partial F_z}{\partial z},$$

$$\nabla \times F = \begin{vmatrix} \dfrac{\hat{R}}{R} & \hat{\phi} & \dfrac{\hat{z}}{R} \\[2mm] \dfrac{\partial}{\partial R} & \dfrac{\partial}{\partial \phi} & \dfrac{\partial}{\partial z} \\[2mm] F_R & RF_\phi & F_z \end{vmatrix},$$

$$\nabla^2 V = \frac{\partial^2 V}{\partial R^2} + \frac{1}{R}\frac{\partial V}{\partial R} + \frac{1}{R^2}\frac{\partial^2 V}{\partial \phi^2} + \frac{\partial^2 V}{\partial z^2},$$

$$\nabla^2 F = \left(\nabla^2 F_R - \frac{2}{R^2}\frac{\partial F_\phi}{\partial \phi} - \frac{F_R}{R^2}\right)\hat{R} + \left(\nabla^2 F_\phi + \frac{2}{R^2}\frac{\partial F_R}{\partial \phi} - \frac{F_\phi}{R^2}\right)\hat{\phi} + \nabla^2 F_z\hat{z}.$$

In spherical coordinates

$$\nabla V = \frac{\partial V}{\partial R}\hat{R} + \frac{1}{R}\frac{\partial V}{\partial \theta}\hat{\theta} + \frac{1}{R\sin\theta}\frac{\partial V}{\partial \phi}\hat{\phi},$$

$$\nabla \cdot F = \frac{1}{R^2}\frac{\partial(R^2 F_R)}{\partial R} + \frac{1}{R\sin\theta}\frac{\partial(F_\theta \sin\theta)}{\partial \theta} + \frac{1}{R\sin\theta}\frac{\partial F_\phi}{\partial \phi},$$

$$\nabla \times F = \begin{vmatrix} \dfrac{\hat{R}}{R^2\sin\theta} & \dfrac{\hat{\theta}}{R\sin\theta} & \dfrac{\hat{\phi}}{R} \\[2mm] \dfrac{\partial}{\partial R} & \dfrac{\partial}{\partial \theta} & \dfrac{\partial}{\partial \phi} \\[2mm] F_R & RF_\theta & R F_\phi \sin\theta \end{vmatrix},$$

$$\nabla^2 V = \frac{1}{R^2}\frac{\partial}{\partial R}\left(R^2\frac{\partial V}{\partial R}\right) + \frac{1}{R^2\sin\theta}\frac{\partial}{\partial \theta}\left(\sin\theta\frac{\partial V}{\partial \theta}\right) + \frac{1}{R^2\sin^2\theta}\frac{\partial^2 V}{\partial \phi^2}.$$

A2.4 Differentiation of vectors

$$\nabla(U+V)=\nabla U+\nabla V$$

$$\nabla\cdot(F+G)=\nabla\cdot F+\nabla\cdot G$$

$$\nabla\times(F+G)=\nabla\times F+\nabla\times G$$

$$\nabla(UV)=U\nabla V+V\nabla U$$

$$\nabla\cdot(VF)=V\nabla\cdot F+F\cdot\nabla V$$

$$\nabla\times(VF)=V\nabla\times F+(\nabla V)\times F$$

$$\nabla\cdot(F\times G)=G\cdot\nabla\times F-F\cdot\nabla\times G$$

$$\nabla\times(F\times G)=(G\cdot\nabla)F-(F\cdot\nabla)G+F\nabla\cdot G-G\nabla\cdot F$$

$$\nabla(F\cdot G)=(G\cdot\nabla)F+(F\cdot\nabla)G+G\times\nabla\times F+F\times\nabla\times G$$

$$\nabla\times\nabla V=0$$

$$\nabla\cdot\nabla\times F=0$$

A2.5 Integral relationships for vectors

$\iiint \nabla\cdot F\,dv=\oint F\cdot ds$ (Gauss's or the divergence theorem)

$\iint \nabla\times F\cdot ds=\oint F\cdot dl$ (Stokes' or the circulation theorem)

$\iiint (\nabla U\cdot\nabla V+U\nabla^2 V)\,dv=\oint U\nabla V\cdot ds$ (Green's first theorem)

$\iiint (U\nabla^2 V-V\nabla^2 U)\,dv=\oint (U\nabla V-V\nabla U)\cdot ds$ (Green's second theorem)

Appendix 3: Physical and materials constants

A3.1 Useful physical constants

Permittivity of free space, $\varepsilon_0 = 8.8542 \times 10^{-12}$ F m^{-1}.

Permeability of free space, $\mu_0 = 4\pi \times 10^{-7}$ H m^{-1}.

Velocity of light in vacuum, $c = 2.998 \times 10^8$ m s^{-1}.

Gravitational constant, $g = 9.81$ m s^{-2}.

Electronic charge, $e = 1.602 \times 10^{-19}$ C.

Electronic mass, $m = 9.109 \times 10^{-31}$ kg.

Charge-to-mass ratio of the electron, $e/m = 1.759 \times 10^{11}$ C kg^{-1}.

Electron volt, $1\text{eV} = 1.602 \times 10^{-19}$ J.

A3.2 Typical material constants

Electric conductivity of copper, $\sigma = 5.8 \times 10^7$ S m^{-1}.

Electric conductivity of aluminium, $\sigma = 3.8 \times 10^7$ S m^{-1}.

Electric conductivity of brass, $\sigma = 1.5 \times 10^7$ S m^{-1}.

Electric conductivity of iron, $\sigma = 1.03 \times 10^7$ S m^{-1}.

Electric conductivity of ferrite, $\sigma = 10^{-2}$ S m^{-1}.

Electric conductivity of fresh water, $\sigma = 10^{-3}$ S m^{-1}.

Relative permittivity of PTFE (Teflon), $\varepsilon_r = 2.1$.

Relative permittivity of paper, $\varepsilon_r = 2$–4.

Relative permittivity of glass, $\varepsilon_r = 4$–7.

Relative permittivity of alcohol (ethyl), $\varepsilon_r = 25$.

Relative permeability of nickel, $\mu_r = 50$.

Relative permeability of cast iron, $\mu_r = 60$.

Relative permeability of ferrite, $\mu_r = 1000$.

Relative permeability of silicon iron, $\mu_r = 4000$.

Relative permeability of Mumetal, $\mu_r = 20\,000$.

Relative permeability of Superalloy, $\mu_r = 100\,000$.

Appendix 4: The SI system

Quantity	Unit	Abbreviation
Length	Metre	m
Mass	Kilogram	kg
Time	Second	s
Electric current	Ampere	A
Temperature	Kelvin	K
Electric charge	Coulomb	C
Electric potential	Volt	V
Permittivity	Farad metre^{-1}	F m^{-1}
Electric field strength	Volt metre^{-1}	V m^{-1}
Volume charge density	Coulomb metre^{-3}	C m^{-3}
Magnetic flux	Weber	Wb
Magnetic flux density	Tesla	T
Permeability	Henry metre^{-1}	H m^{-1}
Magnetic scalar potential	Ampere	A
Magnetic vector potential	Weber metre^{-1}	Wb m^{-1}
Magnetic field strength	Ampere metre^{-1}	A m^{-1}
Resistance	Ohm	Ω
Inductance	Henry	H
Capacitance	Farad	F
Reluctance	Ampere Weber^{-1}	A Wb^{-1}
Force	Newton	N
Energy, work	Joule	J
Power	Watt	W
Torque	Newton metre	N m
Frequency	Hertz	Hz
Radian frequency	Radian second^{-1}	rad s^{-1}

Appendix 5: Bibliography

Introductory textbook on electromagnetism

Hammond, P. (1986). *Electromagnetism for engineers (Third edition)*. Pergamon Press, Oxford.

Advanced textbook on electromagnetic field computation

Binns, K. J., Lawrenson, P. J., and Trowbridge, C. W. (1992). *The analytical and numerical solution of electric and magnetic fields*. John Wiley, New York.

Specialist books on finite differences

Smith, G. D. (1985). *Numerical solution of partial differential equations: finite difference methods (Third edition)*. Oxford University Press, Oxford.
Stoll, R. L. (1974). *The analysis of eddy currents*. Clarendon Press, Oxford.

Specialist books on finite elements

Silvester, P. P. and Ferrari, R. L. (1990). *Finite elements for electrical engineers*. Cambridge University Press, Cambridge.
Lowther, D. A. and Silvester, P. P. (1986). *Computer-aided design in magnetics*. Springer-Verlag, New York.
Chari, M. V. K. and Silvester, P. P. (ed.) (1980). *Finite elements in electric and magnetic field problems*. John Wiley, Chichester.
Zienkiewicz, O. C. and Taylor, R. I. (1991). *The finite element method*, (4th edn), Vols 1 and 2. McGraw-Hill, Maidenhead.

Books and papers on dual energy formulation and the method of tubes and slices

Hammond, P. (1981). *Energy methods in electromagnetism*. Clarendon Press, Oxford.
Sykulski, J. K. (1988). Computer package for calculating electric and magnetic fields exploiting dual energy bounds. *IEE proceedings*, **A135**, (3), 145–150.

Hammond, P. and Baldomir, D. (1988). Dual energy methods in electromagnetism using tubes and slices. *IEE Proceedings*, **A135**, (3), 167–172.

Hammond, P. (1989). Upper and lower bounds in eddy-current calculations. *IEE Proceedings*, **A136**, (4), 207–216.

Baldomir, D. and Hammond, P. (1990). Geometrical formulation for variational electromagnetics. *IEE Proceedings*, **A137**, (6), 321–330.

Appendix 6: A summary of electromagnetic relationships

A6.1 Maxwell's equations

$$\text{curl } \boldsymbol{H} = \boldsymbol{J} + \frac{\partial \boldsymbol{D}}{\partial t},$$

$$\text{curl } \boldsymbol{E} = -\frac{\partial \boldsymbol{B}}{\partial t},$$

$$\text{div } \boldsymbol{D} = \rho_{\text{free}},$$

$$\text{div } \boldsymbol{B} = 0.$$

A6.2 Constitutive equations

$$\boldsymbol{D} = \varepsilon_0 \boldsymbol{E} + \boldsymbol{P} = \varepsilon_0 \varepsilon_r \boldsymbol{E},$$

$$\boldsymbol{B} = \mu_0 \boldsymbol{H} + \boldsymbol{M} = \mu_0 \mu_r \boldsymbol{H},$$

$$\boldsymbol{J} = \sigma \boldsymbol{E}.$$

A6.3 Electrostatics

$$\boldsymbol{E} = -\text{grad } V,$$

$$\nabla^2 V = -\frac{\rho}{\varepsilon} \quad \text{(Poisson's equation)},$$

$$\nabla^2 V = 0 \quad \text{(Laplace's equation)}.$$

A6.4 Magnetostatics

$$\boldsymbol{B} = \text{curl } \boldsymbol{A},$$

$$\text{div } \boldsymbol{A} = 0,$$

$$\nabla^2 \boldsymbol{A} = -\mu \boldsymbol{J},$$

or if there is no current

$$\boldsymbol{H} = -\text{grad } V^*$$

A6.5 Time-varying fields

$$B = \text{curl } A$$

$$\text{div } A = -\frac{1}{c^2}\frac{\partial V}{\partial t}$$

$$E = -\frac{\partial A}{\partial t} - \text{grad } V$$

$$\nabla^2 V - \frac{1}{C^2}\frac{\partial^2 V}{\partial t^2} = -\frac{\rho}{\varepsilon}$$

$$\nabla^2 A - \frac{1}{c^2}\frac{\partial^2 A}{\partial t^2} = -\mu J$$

Index

Ampère, André Marie 43–4
analogue conducting 165
analogue network 165
antenna array 141–3
antenna directivity 141–3
antenna half-wave 139–41
antenna linear 139–41

boundary conditions, electric 33–4
boundary conditions, magnetic 46–8
boundary elements 186–7
bounds, upper and lower 10, 168

capacitance 22–6, 31–2, 145
capacitor 17, 24, 26, 31
Cavendish, Henry 21
cavities, conducting 205, 207
cavities, electric 33
cavities, magnetic 47
charge, bound 30
charge, free 30
charge, moving 94–5
charge, oscillating 134
coenergy, magnetic 219
coercive force 226
computer-aided design (CAD) 188–92
conductance 24
conductivity 2
conductor 1
constants 247
continuity equation 124
coordinate systems 243
Coulomb, Charles Augustin de 21
curl 83
current 1, 41, 44, 134
current element 58–62
current loop 44
cut-off frequency 148

del operator 73
dielectric 31
diffraction 143
diffusion equation 172–4

dimensions 42, 248
dipole, electric 134
dipole, magnetic 44
Dirichlet condition 157
discretization 179
discretization error 170
displacement current 124
divergence operator 75
domain, magnetic 222–4
drift velocity 3, 95

eddy current 109, 225
electrolytic tank 165
electromotive force, motional 94–100
electromotive force, transformer 102–5
electron spin 222
energy, electrokinetic 40–3
energy, heat 1, 40
energy, potential 40–3
energy, stored 17–9, 40
energy, system 26
energy density, electric 27
energy density, magnetic 46
equipotential surface 3
error band 10
error estimation 192

Faraday, Michael 102
Faraday's law 102
field, far 137
field, near 137
field, partial 32
field, rotating magnetic 210–5
field, total 32
field, vector 70
field, vortex 82–5
field stress, electric 27, 64
field stress, magnetic 62–5
finite-difference method 169–74
finite-element method 12, 68, 174–86
fluid, incompressible 21
flux, electric 19–22
flux, magnetic 45, 51–6
flux, mutual 106

flux density, electric 21
flux density, magnetic 45
flux leakage, magnetic 56, 106
flux tube, electric 26–8
flux tube, magnetic 62–5
flux vectors 73–4
fringing flux, electric 26
fringing flux, magnetic 56
focusing, magnetic 101
force, on charged conductor 28
force, on current filament 41

Gauss' theorem 20, 56
generator effect 98
Gilbert, William 43
gradient operator 71–3

Hall voltage 100
Heaviside, Oliver 59
Hertz vector 149
hysteresis 225–6

images 161–4
impedance, reflected 200
inductance 46, 107–9
insulation 3, 30–1
inverse-square law 20–1, 45

Laplace's equation 78–9, 152–5, 169–72,
    174–83
Lorentz condition 133

magnet, permanent 43, 225–6
magnetomotive force 49
Maxwell, James Clerk 6
Maxwell's equations 123–5, 251
Maxwell stress 221
momentum, electrokinetic 117–9
motor effect 99

nabla operator 73
Naumann condition 157

Oersted, Hans Christian 43
Ohm's law 1-2
orthogonality 10

permeability 46
permeance 52, 54–5
penetration depth 113

permittivity 20, 32
polarization 30–2
potential, electric 71–3, 119, 132–4
potential, magnetic vector 86–7, 117–8
potential, magnetic scalar 46, 52
potential, retarded 133–4
potential difference 1, 17
potential gradient 20
Poisson's equation 79
power gain 142
Poynting vector 128–31
proximity effect 209–9

reluctance 52–4
resistance, ohmic 1–3
resistance, radiation 131

saturation, magnetic 54, 204–5
screening, magnetic 158–61
separation of variables 152–8
shell, magnetic 48
skin depth 113
skin effect 109–17
sparsity of equation 183
squares, curvilinear 166–7
surface wave 143
susceptibility, electric 31
susceptibility, magnetic 46

TAS program 227–42
TE and TM waves 148
TEM wave 144, 146
temperature 1
time 92–4
time-constant 29, 108
torque equation 216–8
transformer 105–6
transmission line 145
tubes and slices 3–12, 23–6, 54–5, 167–9

uniqueness 87–9
units 248

variational principle 167–8
vector, axial 81
vector, formulae 243
vector, polar 81
velocity of light 42, 94, 127

wave equation 127
waveguide 147
wavelength 135–9